Geographies of new femininities

Geographies of new femininities

NINA LAURIE, CLAIRE DWYER,
SARAH L. HOLLOWAY AND
FIONA M. SMITH

LONGMAN

Pearson Education Limited
Edinburgh Gate
Harlow
Essex CM20 2JE, England

Published in the United States of America
by Pearson Education Limited, New York

First published 1999

ISBN 0 582 32024-0

British Library Cataloguing-in-Publication Data

A catalogue record for this book is available from the British Library.

Set by 35 in 10/12pt Sabon
Produced by Addison Wesley Longman Singapore (Pte) Ltd.
Printed in Singapore

Contents

List of figures   *page* viii
List of tables   ix
About the authors   x
Preface   xi
Acknowledgements   xiv

1 *Introduction: geographies and new femininities*   1
   Introduction   1
   What are 'femininities'?   3
   Geographies and femininities   11
   Outline of the book   14

2 *Changing worlds? Changing femininities?*   16
   Introduction   16
   Globalising economies, cultures and politics   17
   Fractured and fracturing identities   26
   Spaces of oppression or spaces of opportunity? The
   emergence of new femininities   32
   Conclusion   40

3 *Working with genders and geographies*   41
   Introduction   41
   The research projects   43
   Working in 'the field' with power and positionality   47
   Working with 'gender'   53
   Working through geographies   59
   Conclusions   65

4 *The shifting geographies of femininity and emergency
   work in Peru*   67
   NINA LAURIE
   Introduction   67
   Crisis and emergency employment in Peru   67
   PAIT: new spaces for women?   71
   Multiple and fractured identities   74
   Shifting the ground of public/private and home and work   83
   Superseding boundaries   88
   Conclusion   89

5 *Reproducing motherhood*   91
   SARAH L. HOLLOWAY
   Introduction   91
   The changing context of mothering in Britain   92
   The Sheffield case study   94
   Mapping mothering   96
   Configuring geographies and new femininities   108
   Conclusion   112

6 *Contested territories: women's neighbourhood activism
   and German reunification*   113
   FIONA M. SMITH
   Introduction   113
   Problematic relations between eastern German women and
   politics   114
   The case study: neighbourhood activism in Leipzig   117
   Learning the ropes: women's changing political subjectivities in
   neighbourhood action   119
   Reworking local political femininities and their geographies   127
   Conclusion   134

7 *Negotiations of femininity and identity for young
   British Muslim women*   135
   CLAIRE DWYER
   Introduction   135
   Theorising about young women and femininity   137
   The research context   138
   Debating 'appropriate' femininities   138
   Negotiating 'appropriate' femininities   143
   Constructing alternative Muslim femininities   146
   Negotiations of femininity and identity: imagining futures   148
   Conclusion   151

8 *Configuring and reconfiguring geographies*   153
   Introduction   153
   Geographical constitution of femininities   154
   Reconfiguring change   173
   Conclusion   185

9 *New femininities*   187
   Introduction   187
   New femininities?   188
   Reworking femininities   192
   The geographies of new femininities   195

   Further reading   198
   Emergency work and 'workfare'   198
   Peru: political economy   198
   Gender identities in Latin America   199
   Motherhood and mothering   199
   Childcare   200
   Women in eastern Europe and the post-Soviet states   200
   Women in the GDR and German reunification   201
   German reunification   201
   Femininity and adolescent girls   201
   The theorisation of 'new ethnicities'   202
   Islam in Britain   202
   Identities of young British South Asian women   202
   Debates about the veil   202

   Bibliography   203
   Index   221

List of figures

1.1 A powerful image: cult cartoon figure Tank Girl 'fuels momentum to assertiveness'   2
1.2 Domesticity and the Empire: an Empire Cook Book   6
1.3 'All women are courageous, strong, beautiful!' Feminist campaigning in eastern Germany. Campaign poster for the Independent Women's Association with the slogan 'a carefree future for women and children in our European home'   9
1.4 Advert selling contact lenses to change the colour of eyes from brown to blue   11
2.1 New ways of teaching about divisions in the world   17
2.2 Women as the 'Losers of Unification'   38
3.1 Neighbourhood action group information stand at a local festival, Leipzig, 1992   46
3.2 Map of Lima showing Nina Laurie's three case study sites – José Galvez, La Tablada and Villa María   62
3.3 Map of Andahuaylas showing Nina Laurie's case study site   62
4.1 Defaced propaganda graffiti for PAIT, La Tablada. It shows PAIT's name crossed out and replaced with COOPOP, which was what the programme was called for the first two years of Fujimori's regime (1990–1992)   68
4.2 Changing decision-making patterns concerning women's and men's use of public space   84
6.1 Creating public space through a local street festival, Gohlis, 1992   119
6.2 A children's arts and environment project staffed by women on the ABM programme, Plagwitz, 1992   131

List of tables

5.1 Socio-economic statistics for Hallam and Southey Green, 1991
Census   95
5.2 Socio-economic composition of households with pre-school aged
children in Hallam and Southey Green   95

About the authors

NINA LAURIE is a lecturer in development studies in the Department of Geography at Newcastle University, UK. She has worked extensively in the Andean region of Latin America and works collaboratively with colleagues in gender studies centres at the University of San Simón Cochabamba, Bolivia, the Catholic University of Peru in Lima and the University of Chile in Santiago, as well as with feminist geographers in the UK. She is currently working on transnational indigenous communities in the Andes with Sarah Radcliffe (Cambridge University).

CLAIRE DWYER is a lecturer in geography at University College London, UK. Her research interests are in the geographies of gender and ethnicity and she has published work on the politics of state-funded Muslim schools in the UK and the Netherlands and the identities of young Muslim women. Her current research focuses on British–Asian businesses and transnational communities.

SARAH L. HOLLOWAY is a lecturer in human geography in the Department of Geography at Loughborough University, UK. Her research interests lie in the fields of feminist geographies, social geography and young people's geographies. She has published work on mothering cultures and the use of pre-school child-care provision in the UK. Her current work (with Gill Valentine, University of Sheffield) focuses on children's experiences of computer-mediated communication, and looks in particular at the ways in which children's technological competencies on the Internet are shaping their use of time, domestic relationships, friendship networks and sense of 'community' from the local to the global scale.

FIONA M. SMITH is a lecturer in human geography at the University of Dundee, UK. Her principal interests lie in the areas of political, urban and feminist geography. These interests combine in her research on the geographies of urban activism in post-communist 'transitions'. In particular this work has focused on eastern German responses to and actions surrounding reunification. She is currently engaged in a project examining the gendered politics of economic and urban restructuring in eastern Germany.

All four authors are active members of the Women and Geography Study Group of the Royal Geographical Society with the Institute of British Geographers.

Preface

The origins of this book lie in our respective doctoral research projects and in our subsequent studies which have included examinations of a range of perspectives on changing gender identities and their geographies. Bringing together primary research material from four distinctive studies and involving collectively and singly authored work, this book aims to forward debates on the changing geographies of women's lives and identities – what we have called 'the geographies of new femininities'.

The actual process of writing a text such as this often receives relatively little consideration in geography and, although some texts have begun to explore these issues, it is common for readers to be presented with a finished product which gives little indication of the processes which go on in the production of such 'knowledge'. Each of us has researched a range of projects in academic geography for several years and has her particular specialism in a quite distinctive area of study, often focusing in different geographical locations. However, Nina Laurie developed the idea that combining these disparate studies could produce new insights on issues of gender and geography. In writing this book we began with the four individually authored chapters on particular case studies and from that began building the surrounding chapters which are all collectively authored. Each also commented on subsequent revisions to the individually authored chapters.

Such collaborative authoring required its own geographies – meetings at various times in Nottingham, London, Newcastle and Dundee provided for intensive collective working as homes and weekends became the spaces and times of work, but also of hospitality and friendship. The joys and frustrations of e-mail and telephone conferencing provided key means for circulating chapters and refining drafts while we worked from a series of locations in the UK, Europe, Southern Africa and South and North America. For all of us, these collective spaces provided some of the most fulfilling and productive elements of writing the book (and the most enjoyable!) as we came to create something that is, we believe, more than the sum of its parts. The book is an ongoing engagement between and around our individual topics of concern and has produced insights that do not come from one approach alone. Comparing and contrasting multiple perspectives on issues has resulted in what we hope is a book that combines detailed knowledge of the workings-out of the geographies of new femininities in particular spaces and places with a broader exploration of the ways that these examples can illuminate key debates and processes involved in globalisation and the negotiation of identities.

Not only has the writing of this book involved a particular geography, it has also involved a critical engagement with the contemporary political economy of academic research and publishing. This book is shaped by processes that led to the production of the research on which it is primarily based (access to free education, scholarships, freedom to study and to conduct fieldwork, etc.) and by the value placed on various forms of writing by the current regulations for research quality audit at universities across the UK in which we all work. This audit, or Research Assessment Exercise (RAE), has a 'hierarchy' of publishing formats. In the discipline of geography this means that refereed journal articles are valued above almost any other form of publication. They are seen as 'better' than book chapters in edited volumes and much better than student texts, which are not usually considered to be research focused. Furthermore, singly authored work is most highly prized. These assumptions are embedded in the RAE and promotions exercises alike and are clearly evident when people apply for new jobs or research funding.

In writing this book in the way that we have, we chose to question some of these values. Having all been involved in the collective of members of the Women in Geography Study Group (of the Royal Geographical Society with the Institute of British Geographers) which came together to write the textbook *Feminist Geographies: Explorations in Diversity and Difference* (WGSG, 1997) we were already convinced of the strengths of collective forms of working. This book, and other examples of collectively authored texts, moves beyond the bounds of work produced as 'lone scholarship', with its association with thinking in splendid isolation, to develop work rich in dialogue and productive connections. (Actually we would argue that lone scholarship rarely occurs since knowledge is fundamentally produced in social settings.) This is not to argue that collective writing is superior to singly authored work, since that would simply replace one hierarchy with another. Rather we seek to assert that such texts are equally valid and often have something quite distinctive to offer.

Indeed, we hope to show that collective authoring can move beyond the frustrations of edited collections where a series of chapters are 'topped and tailed' with an introduction and conclusion which set the scene and summarise findings but often lack the space fully to explore the relationships between the different material presented and the importance of the similarities and differences identified. As a challenge to the distinction between 'research' and 'student' texts, our aim in the production of this book is to 'demystify' research-based journal writing directed at a select audience of 'experts'. We aim to make our research more accessible to peers and students alike in practical ways by taking into consideration the language and style of writing and the limits that many library budgets place on journal acquisitions. Our book is research-based and is a text written for what we hope will be a wide and varied audience. Furthermore, we see this as one book among many engaging with these issues and would encourage readers to make use of the further reading sections found at the end of the book to explore and develop interests and debates which this book sparks off for them.

Nina Laurie, Claire Dwyer,
Sarah L. Holloway and Fiona M. Smith
Newcastle, London, Loughborough and Dundee

Acknowledgements

In producing this book, we have received indispensable help and support from a number of people whom we wish to thank both individually and collectively. Matthew Smith, our editor at Addison Wesley Longman, has been untiring in his support for this project. Without him the book would probably never have come to fruition. Many colleagues and friends have provided invaluable comments and suggestions on our ideas and we would specifically like to acknowledge the generous support received from Audrey Kobayashi, Doreen Massey, Chris Philo and Gill Valentine. Each of us individually also acknowledges our supervisors for the doctoral research on which the book is based: Ann Varley; Peter Jackson and Jacquie Burgess; Nicky Gregson, Charles Pattie and Paul White; and Ronan Paddison, respectively. The research was facilitated by the following scholarships: Economic and Social Research Council awards R00429224177, R00429234082, R00429134137; and the Carnegie Trust for the Universities of Scotland, respectively. For their help in producing this book we would also like to thank the following people: Ergul Ergun for her work in compiling and checking the bibliography; Kathryn Morris Roberts for providing the images for Figure 1.1; Ann Rooke for her work in drawing maps and compiling Figure 1.1; Irene Dunn for the image in Figure 1.2; Grit Beck for material on women in Germany; Morag Bell and Jo Bullard for commenting on sections of Chapter 1; and Irith Williams, our production editor.

The authors are very grateful for permission to reproduce the following copyright material: Figure 1.1 reproduced courtesy of Deadline Publications, Figure 1.3 reproduced courtesy of Haus der Geschichte der Bundesrepublik Deutschland, Bonn, Figure 2.1 reproduced courtesy of United Feature Syndicate, Inc., Figure 2.2 reproduced courtesy of Marie Marcks, Heidelberg, 1992, 'Alice's Song' reproduced courtesy of Ricky Ross.

Finally, as we mentioned previously, the genesis of the book involved meetings in various homes and academic departments and we thank in particular Chris, Paul, Ben, Sarah and Sabine for putting up with our 'invasions' and the time spent working. Above all, however, we want to acknowledge the collegiality, support and friendship which we have found in working with each other in this project.

CHAPTER ONE

Introduction: geographies and new femininities

Introduction

The phenomenon of new femininities has been much discussed in the (British) media and popular press in recent years. Research evidence suggests that girls are out-performing boys at all levels of school education (Wilkinson and Siann, 1995), and are advancing in the labour market in traditionally 'masculine' spheres such as the law and medicine. This has led to claims that 'everywhere you look, you see individual women who are freer and more powerful than women have ever been before' (Walter, 1998: 1). Moreover, the celebration of icons of new femininity in popular culture – such as the 'girl power' of the British band the 'Spice Girls', the comic strip heroine 'Tank Girl' (see Figure 1.1), and the emergence of girl gangs or Ragga girls – is leading cultural theorists to suggest that 'new modes of femininity' (McRobbie, 1993) are emerging for young women which are 'both excessively and transgressively feminine' (McRobbie, 1996: 36). The idea of hyped femininity, of 'ultra-woman', is still evident when these new femininities are dissociated from heterosexuality, most notably through ideas of the 'lipstick lesbian' and the focus on lesbian chic in the mainstream press in the early 1990s (Bell *et al.*, 1994). Such celebrations of new femininity may be challenged by the realisation that gender inequalities remain embedded within economic and social structures. For example, women who have children still find themselves worse off than their male counterparts and women without children (Benn, 1998; Walter, 1998), and the 'feminisation' of poverty continues to increase. This evidence has led some critics to argue that we need a 'new feminism' that will address such inequalities while capitalising on 'the new female confidence that exists in Britain' (Walter, 1998: 4).

While we are interested in this celebration of new femininities in Britain, and particularly in the ways in which it influences the lives of young women growing up in a multi-cultural society (see Chapter 7), we take a broader and less Anglo-centric view of the notion of new femininities. We see changing gender identities and the emergence of an interest in the geographies of femininities as stemming from two important dimensions within geographical

Figure 1.1 A powerful image: cult cartoon figure Tank Girl 'fuels momentum to assertiveness'. Reproduced courtesy of Deadline Publications

research. The first is a focus on the ways in which processes of globalisation, such as economic and welfare restructuring, the 'new world order' and the implications of new information flows and technologies, may be shaping new employment and political roles for women. Economic, cultural and political processes are working in different ways across the globe and may result in new forms of social relations which work to transform gender identities and open up new spaces for engaging with the construction and contestation of femininities. Rather than creating greater opportunities for women, processes of globalisation may also be associated with the remaking or reworking of existing gender divisions.

The second important dimension resulting in an increased focus on the geographies of femininities is the influence of theoretical shifts in the study of gender identities. In particular, feminist geographers are increasingly interested in the fluidity and multiplicity of identities and in the fracturing of identity formation. This interest emerges, in part, from the changing gender politics of everyday lives. It is also the result of an engagement by feminist geographers with a range of different theoretical literatures on identity including postcolonial theory and the 'new ethnicities' literature; work by feminists and others on sexuality and the body; as well as post-structuralist theories of subjectivity and identity. Thus a combination of social, economic, political and cultural processes, alongside theoretical shifts in the ways in which we understand them, have shaped feminist geography and have produced a focus on the geographies of femininities.

In the remainder of this chapter we sketch out further what we mean by geographies and new femininities and offer an outline for the rest of the book. We begin by defining the term femininity and describing how feminist geographers have become increasingly interested in the analysis of different kinds of femininities. We then consider the relationship between geographies and femininities, tracing the intersection between the construction and contestation of different femininities through different spatial contexts and in relation to different geographical concepts. Finally, we provide an overview of the structure for the book as a whole.

What are 'femininities'?

First, what do we mean by 'femininity' and 'femininities'? A starting point might be to define femininity in relation to the social construction of gender. While sex – male or female – might be understood as a category based on biological difference, gender is understood as a social construction organised around biological sex. Thus, individuals are born male or female but, over time, they acquire a gender identity, that is an understanding of what it means to be a man or woman (WGSG, 1997: 53). This gender identity is defined as masculinity or femininity. These gender identities are often naturalised, relying on the notion of biological difference, so that 'natural' femininity encompasses, for example, motherhood, being nurturing, a desire for pretty clothes and the exhibition of emotions. However, the social relations between genders must be recognised as socially constituted and historically and geographically differentiated. As Connell (1995) argues, hegemonic gender relations become established at particular periods, and in particular contexts, reflecting the dominance of specific ways of being masculine or feminine. For example, the dominant notion of appropriate masculinity has been fixed at different points as being a breadwinner, and femininity as being a mother while masculinity and femininity have very often both been understood within the context of hegemonic heterosexuality. This definition makes it clear that there are multiple femininities and masculinities but that in different contexts particular masculinities and femininities become hegemonic or 'normal'. These normalising processes often result in sanctions for those women or men who are deemed to be exhibiting 'deviant' or 'inappropriate' femininities or masculinities.

As this definition suggests, femininity cannot be understood as a single gender identity and instead it is more useful to think of multiple or fractured femininities. A focus on femininities reflects the instabilities and contradictions of the social construction of femininity within particular times and places. It also recognises that gender identities intersect with other aspects of identity – such as class, 'race', disability, sexuality, age and so on – resulting in the multiple constitution of gendered identities with other socially constructed identities. As a result, femininities are also classed, racialised, sexualised as well as being geographically and historically specific. This recognition of a multiplicity of femininities suggests that while hegemonic femininities often draw upon 'natural' or 'essential' associations between the biological, sexed body and gender identities, these associations are not straightforward. Instead we might consider that sex and gender are instead dialectically interrelated (Jaggar, 1984; Butler, 1993) and that our biological and social selves are mutually constituted. Thus, drawing on Connell (1983, cited in Jackson, 1991: 208) we can see the embedding of femininity within the body as a lengthy and complex social process, a historical rather than a biological fact.

This argument suggests that it is possible to think about femininities as multiple and fractured, but also as socially, historically and geographically constituted. Individual 'biological' women are involved in the construction and contestation of femininity, through different means and drawing on different repertoires or resources. This is then *performed* in the contexts of individuals' daily lives. Yet the carrying out or acting out of particular sets of ideas about how women are to behave and be is not as voluntaristic as the term 'performance' might suggest. The lives of individuals are clearly constructed within economic, political and social structures through which dominant femininities (and masculinities) are determined. Individuals are thus positioned, and position themselves, in particular ways in relation to dominant discourses and practices of gender identities. Individually, or collectively, they may be complicit, subverting and/or critical (Cocks, 1989). They may also locate themselves or be located in relation to other forms of identification, not simply gender identities.

These abstract ideas can be best illustrated through particular examples which show how ideas about femininity have been constructed and contested within different historical periods and spatial contexts. We have chosen three examples which relate to the four main case studies in this book. The first examines the construction and contestation of femininities within a British, or more specifically *English,* context. This analysis focuses on the lasting influence of the mutual constitution of class and gender at different points in English history: first at the end of the eighteenth century; second during the inter-war period when notions of empire also influenced understandings of identity; third in the immediate post-war and contemporary periods. These ideas about femininity are important to the analyses elaborated in Chapters 5 and 7 which deal with concepts of mothering in the UK and the construction of identities among young British Muslim women respectively. The second example focuses on constructions of femininities in Germany in the pre-war, post-war and reunification periods. This not only provides a useful comparison with UK constructions of gender at these times but also illustrates, in particular, how ideological shifts at the level of the nation-state can influence changing gender regimes. Furthermore, it provides a historical background to gender identities in 'transition' in eastern Germany which are explored in Chapter 6. The final example also focuses on the role of political ideologies in the construction of femininity. The focus here is on the ways in which understandings of femininity are configured in relation to discourses of nationhood in Latin America. While this analysis is important for the discussion of changing femininities in Peru in Chapter 4, it also illustrates the significant role that the state and rhetorics of nationhood play more widely in constructions of gender identities (see Sharp, 1996; Yuval Davies, 1997).

A useful starting point for considering different constructions of femininity in the British context is the work of Leonore Davidoff and Catherine Hall on the lives of middle-class men and women (Davidoff and Hall, 1987). Davidoff and Hall argue that gender played a central role in the construction of middle-class values, family life and property relations in late-eighteenth century and early-nineteenth century England. Middle-class life was structured around an assumed 'moral order' through which appropriate masculinity and femininity was based on Christian ideology. Central to this ideology was an assumption of separate spheres – the 'public' and the 'private' sphere – with women being 'naturally' associated with the home or 'private' sphere. Yet, while this association was naturalised, women's role within family enterprise (and so in support of the capitalist cause) was still fully recognised. Davidoff and Hall stress that neither femininity, nor 'manliness', were fixed categories but instead 'were constantly being tested, challenged and re-worked both in imagination and in the encounters of daily life' (Davidoff and Hall, 1987: 405). This tension is illustrated by evidence that men systematically felt the need to contain women. Thus, as new forms of capital gave women increasing financial independence, this was curbed through property laws, laws of inheritance and new forms of economic organisation which intersected with definitions of femininity (Davidoff and Hall, 1987: 415). The central theme of Davidoff and Hall's argument is not simply that masculinities and femininities are never fixed and are constantly in process, but also that gender and class always operate together.

Figure 1.2   Domesticity and the Empire: an Empire Cook Book

Foreword by Mrs Baldwin

It is not given to us all to be Empire-builders, but there is one way in which we can very definitely help Empire Trade, and that is by buying British and Empire goods. We depend very largely on our trade with the Overseas Dominions and Colonies, and they help us by giving a preference on British-made goods. Is it not better that the money we spend on food should go to help our kinsmen overseas who are planting and growing food for us, rather than to foreigners who have no interests in the British Empire? Not only are British goods better and cleaner than foreign goods, but they are generally cheaper.

A further example of the relationship between constructions of femininity and class in the English historical context is provided by Alison Light's exploration of the ways in which ideas about national identity during the inter-war period were bound up with notions of femininity (Light, 1991). Light highlights how anxieties about both national identities and femininities characterised the post-World War I period. She traces the makings of an Englishness which draws upon a middle-class femininity and particularly the centrality of the home and the domestic sphere. This sphere is seen as 'modern', providing the sense of a new culture of 'homemaking', but is also deeply conservative. These tensions are carried through in the examples of the construction of contemporary femininities in the UK as discussed in Chapters 5 and 7. This making of a middle-class Englishness during the inter-war period also ties together ideas of Empire and femininity where the home and 'feminine integrity' and women's command of domestic space were integral to the success of imperial interests (see Figure 1.2). Thus, gendered and classed femininities are also constituted through ideas about imperialism and whiteness. As Vron Ware (1992) illustrates in her book about Englishwomen and Empire, constructions of femininities cannot be separated from the racialisation of identity.

Light's (1991) work emphasises the extent to which particular historical moments have been characterised by anxieties about 'appropriate' femininities. Clearly, the inter-war period in Britain was one such period when confusion about national identity became articulated through a focus on the role of women. The same anxieties were also evident during the 1950s, the post-World War II period, when an intense debate focused upon the need for women to abandon the workplace and return to 'traditional' family values. As Deborah Leslie (1993) illustrates, this debate was played out in particular through advertising which attempted to redefine women's role in accordance with a new industrial order (Ewen, 1976). Leslie traces the same debates in what she defines as the 'new traditionalism' embodied in contemporary advertising and editorials in women-oriented magazines in the UK and the US such as *Family Circle* and *Good Housekeeping*. In this analysis she suggests that the 1990s might again be recognised as a period of anxiety over 'appropriate' gender identities. She argues that the discourses of contemporary magazines which articulate a 'new traditionalism' founded on domesticity, tradition and familialism, can be seen as a response to changing social and economic conditions – such as the breakdown of 'traditional' families, changing gendered work patterns, a crisis in patriarchy – which seeks to reaffirm a nostalgic and 'traditional' version of femininity. In making this argument she draws upon the work of Linda McDowell (1991) who examines the gendering of post-Fordism and the links between an ideology of domesticity and idealised femininity and a flexible, part-time female labour force (this argument is discussed in more detail in Chapter 2). While Leslie analyses some of the tensions within contemporary constructions of femininity and the rise of a new traditionalism within the context of US right-wing politics, current British debates – particularly around motherhood and the demonisation of 'single mothers' (Laws, 1994; McRobbie, 1994) – can also be seen to provoke similar tensions around constructions of femininity.

These issues of women and home, and women and work also constitute key poles of debate and struggle around 'appropriate' femininities at different points in German history. A recent exhibition in Germany, entitled 'Unequal Sisters? Women in East and West Germany' (Schäfer, 1997) outlined the common and divided histories of German women. A common history is found in the rise of women's emancipation in the post-World War I years where, after an initial reduction of female labour-force participation following demobilisation, women increasingly entered paid employment, including the professions. Universal suffrage, which in 1919 saw 41 women parliamentarians elected, and widening access to education contributed to the sense of an emerging 'new woman' (Frevert, 1997). The 'new woman' also looked different, with a change in body image. The short hair, androgynous appearance and emphasis on physical fitness represented by the women in the photograph from 1928 on the cover of this book marked a radical shift away from 'traditional' femininity. Such change fell far short of giving women equality with men. For example, virtually all women stopped work when they married. Nevertheless, echoing attitudes found in Light's

(1991) work on England at the time, 'many thought "emancipation", as it was called, was going too far, eroding the foundations of state and society' (Frevert, 1997: 12). The conservative, nationalist backlash of the National Socialist period (1933–45) advocated the 'return' of women to the roles of wife and mother, roles that were central to the regime's racial politics of developing a 'pure', 'Aryan' population. Protecting women's roles meant that in World War II, unlike Britain, the German government refused to enrol women on the 'home front' in armaments or other industrial production, 'preferring foreign forced labour, male and female' (Frevert, 1997: 13). In the aftermath of the destruction of World War II, however, women, who made up the majority of the population, were central to reconstruction work, with armies of *'Trümmerfrauen'* (rubble women) clearing ruins and debris. After this period considerable divergence in approriate femininities emerged with the establishment of separate western and eastern German states in 1949.

Processes in the Federal German Republic (FGR; western Germany) parallel in many ways those discussed by Leslie (1993) for the USA and England. The 'economic miracle' of the 1950s provided full male employment while wage levels and social service provision were organised around the 'male breadwinner/female housewife' model. Over time the rise of second-wave feminism, increases in female labour-force participation, falling birth rates, later marriage, increased cohabitation and divorce were all similar to trends found across many western countries over this period (Kolinsky, 1993). However, female labour-force participation rates remained below those of many other western European countries and structures of childcare and schooling provision, for example, still operated on the assumption of a mother at home for much of the day (although this has begun to change in recent years).

In contrast, the government of the German Democratic Republic (GDR; eastern Germany) set out to replace bourgeois ideals with a socialist ideal of the working woman: 'real equality for [a] woman is only present when she has learned a trade and is in the position to carry out a job of real value to society' (Walter Ulbricht, first GDR State Premier, speaking in 1949; quoted in Didczuneit, 1997: 17). This also involved a deliberate distancing from western norms. In 1959 a GDR women's magazine printed a quiz from a western German publication which asked 'are you attractive to men?' It then contrasted the 'ideal woman' of the West ('from a good home, a good homemaker, educated, young and beautiful, and with savings') with answers on the attractive qualities of women in socialism ('intelligence, cultural interest, the self-confidence of a working woman independently going her own way') (Didczuneit, 1997: 23).

By the 1980s virtually all women in the GDR did work, and virtually all combined this with having children. As in other communist states, state systems of childcare provision, family benefits and other social services were designed to allow women to combine family and work, but the lack of emphasis on changing masculine roles left women carrying a multiple 'burden'. The resultant self-image was ambivalent for many women who recognised the exertion required to fulfil these roles and the continuing discrimination in access to senior jobs and particular sectors (Nickel, 1990), but also knew their own important contribution to household income and their strength and abilities in juggling

Alle Frauen sind mutig! stark! schön!

Unabhängiger Frauenverband - Auch für Frauen und Kinder eine sorgenfreie Zukunft in dem Europäischen Haus

Figure 1.3 'All women are courageous, strong, beautiful!' Feminist campaigning in eastern Germany. Campaign poster for the Independent Women's Association with the slogan 'a carefree future for women and children in our European home'. © Haus der Geschichte der Bundesrepublik Deutschland, Bonn

multiple demands (see Figure 1.3). For most, having full-time employment *and* family was the 'norm': 'We had a husband and children, and we worked. What else would you expect?' (Hildebrandt, 1996: 39). Notably these femininities did not include the possibilities of non-heterosexual femininities (Schenk, 1993).

Fuelled by the selective images of GDR women common in the West (drug-enhanced, masculinised bodies of sportswomen or unstylish, down-trodden working women), there was considerable surprise in the West post-reunification that many women did not wholly reject this 'double burden' in favour of western luxury and a 'hyped' femininity (although this is also found). Some of the difficulties in bringing together experiences and expectations from eastern and western Germany in the context of German reunification are explored in Chapter 6. Moreover, it is clear from this example both that no simple, universal, linear progression exists in the ideas and practices of what constitutes femininity and that gender identities are fractured by divisions such as those around 'race', nationality, sexuality or stage in the life course.

Having provided brief examples of how debates about 'appropriate' femininities have been played out historically within the British and German contexts, we now want to draw on the third example which focuses on appropriate femininities (and masculinities) in Latin America. In their recent book, Sarah Radcliffe and Sallie Westwood (Radcliffe and Westwood, 1996) illustrate the ways in which ideas about nationalism and associated modernist discourses of democracy in Latin America have been filtered through ideologies about gender. They explore how nationalisms were constructed through imagined brotherhoods, while women were included in the project of nation-building in specifically 'feminine' roles – as reproducers, as social agents in community development and as domestic crisis management agents. In particular historical moments women were called upon to play a role in nationalist struggles yet, as Radcliffe and Westwood (1996: 136) emphasise, these roles were articulated through the construction of specific femininities. Thus, in Argentina in 1949, for example, the Partido Peronista Femenino (PPF: Peronist Feminine Party) encouraged women to participate in populist nationalism through a discourse which saw women as feminising social action while remaining subordinate to their husbands. This role was personified in Evita Perón who became the embodiment of a femininity based on social action and yet played a secondary role to her husband, the dictator Juan Perón. She is famed for saying that in their relationship 'He is the figure, I the shadow' (Fisher, 1993: 153).

Gender and 'race' are also linked together in the construction of national identities within the Latin American context just as they have been in the case of English national identities discussed earlier. For example, Radcliffe and Westwood (1996: 139) emphasise the extent to which whiteness is prized and linked particularly to the elite and upper-middle-class urban social groups in a nation such as Ecuador. Indigenous women, from highland Andean communities, not only internalise their 'non-white' features as 'ugly' (Weismantel, 1988) but also remain excluded from discourses of femininity which focus upon certain racialised characteristics as symbolic of the imagined national community. Hence advertising images in the Andean region such as those shown in Figure 1.4 promote 'white features' and 'blue' eyes over the 'brown' eyes and skin associated with indigenous and '*mestizo*' (mixed 'race') Latin Americans.

The role of religion in framing notions of appropriate femininity has been highlighted earlier in the English example where ideals of femininity – particularly the importance of the domestic sphere – were reinforced through an appeal to a 'natural' order associated with Christianity. The same appeal to religious beliefs is evident within Latin American debates about appropriate femininities. Here the male culture of '*machismo*' is counterposed to '*marianismo*', an ideology of female moral superiority which draws upon the Mediterranean and Catholic colonial heritage. *Marianismo* (literally 'the cult of Mary') draws upon the contradictory representations of Mary the 'virgin mother' of Jesus. It evokes both maternal values and ideas of shame and honour so that chastity, meekness and self-sacrifice are seen as desirable feminine traits (Stevens, 1973; Fisher, 1993). As well as informing the case study on Peru in Chapter 4, this example raises the wider question of how religion operates as an element in the negotiation of femininities (see Chapter 7).

Through these three brief examples it is possible to see the ways in which ideas about appropriate femininities are produced within different places and within different historical contexts. It is also evident that differences between women, for example in terms of class, 'race' and religion, are important in the construction of femininities.

Figure 1.4 Advert selling contact lenses to change the colour of eyes from brown to blue.
© Nina Laurie

Geographies and femininities

Having illustrated what we mean by femininities, in this section we want to raise a series of questions concerning geographies and femininities. Feminist geography, as we suggested at the beginning of this chapter, has increasingly focused on femininities and geographies (with a growing amount of work also looking at the geographies of masculinities: Jackson, 1991; Massey, 1995; Shepherd, 1998). This focus reflects a shift within feminist geography, highlighted in many different accounts (McDowell, 1993; WGSG, 1997), from work that focused on the 'geography of women', via the study of 'gender roles' and subsequently 'gender relations', to a consideration of the geographies of gender identities. As McDowell (1993: 171) argues, geographers are now asking 'more complex theoretical questions about the extent of variety in the spatial constitution of gender and the specific ways in which the characteristics of masculinity and femininity vary between spaces, classes and ethnicities'. As we outline in Chapter 2, this shift reflects both theoretical and material changes. In terms of feminist theory, feminist geographers have increasingly grappled with issues of difference between women. Questions of difference have emerged partially in response to the writing of black women and women of colour who have critiqued the unitary category of (white) women presented in feminist analysis. Post-structuralist work has also suggested theorisations of identity as multiple and fractured and critiqued the reliance on the body as a site of (essentialist) sexual difference. While we explore these issues in more depth in Chapter 2, here we want to draw on the work of feminist geographers to suggest how we might develop a three-fold typology of the ways in which geographies and femininities are mutually constituted.

First, feminist geographers have suggested that we need to think about the difference place makes (Massey and Jess, 1995) in terms of geographical context. In other words the ways in which femininities are forged differently within different countries, regions or more local places. 'Place' not only shapes 'gender' but femininities in turn are part of what shapes those places. Second, work on gender identities has also started to explore how femininities are forged in and through particular spaces, sites and networks. Thus femininities might be constituted differently within particular types of spaces, such as work spaces or domestic spaces, and may vary between different work or domestic environments. Third, the spaces and places involved in the two previous points are also constituted through cultural discourses about gender, so that the 'home' for example, becomes encoded as a 'feminine' space. There is therefore a dialectical process through which gendered discourses and geographical categories are mutually constituted. Below, we consider each of these points in more detail.

Feminist geographers have long since highlighted differences in gender relations across regional and national space. They have argued that femininities are therefore constituted differently in different places, and that they in turn help to make those places what they are. Work by McDowell and Massey (1984), for example, shows how regional economic variations and different gender relations develop over time in ways which draw on each other. For example, they show how the north-east of England, which had historically been characterised by a highly segregated gender division of labour where women worked only in the home, attracted branch-plant manufacturing in the post-1970s economic restructuring partly because the women of the region were seen as a green labour-force who would be docile and could be paid low wages. In contrast, women in the north-west of England, who had a history of employment outside the home, were more likely to be unionised and demand higher wages. This, along with a lack of regional aid, made north-west England less attractive to capital. These two regions were characterised by different gender relations, shaping the femininities of women in the two areas in different ways, and these gender relations were an important part of the geographies of those places, for example in attracting different levels of investment to the regions.

Today we can see how femininities are constructed and appropriated by capital within many different geographical spheres – work on the employment of women in Newly Industrialising Countries (NICs), or Free Trade Zones (FTZs), for example, highlights how women are prized as workers because of assumed feminine characteristics, such as dexterity, submissiveness and docility (Elson and Pearson, 1981; Stichter and Parpart, 1990; Arregui and Baez, 1991; Rosa, 1994). More recent work on the expansion of telemarketing industries and the emergence of new call centres as key employers in areas such as the north-east of England highlights similar issues. The availability of a cheap work-force with assumed feminine characteristics still plays an important role in location decisions and therefore in the experiences of women in this area. For example, in the development of call centre employment, the 'emotional labour' and the 'smile down the phone' factor provided by a female work-force reflect particular constructions of femininity as employers see women as 'more natural' on the telephone (Marshall and Richardson, 1996: 1855).

These places which 'make a difference' are not simply bounded locations. The varied connections between places suggest that the definition of places should incorporate an awareness of their interconnectedness: 'at the same time as recognising the individual uniqueness of each, places may be conceptualised as essentially open and porous, as interlinked' (Massey and Jess, 1995: 67). In this 'progressive sense of place' (Massey, 1994) places are constituted by, and in turn constitute other places, and connect in various ways to wider processes. For example, those working in call centres in north-east England and in factories in FTZs in South East Asia may work for the same multinational corporation, whose employment policies may both shape these places and be shaped by the processes within these places. Places are not 'fixed' but rather constitute and are constituted through shifting 'global/local' connections. These connections are fundamentally unequal and result in and from differential access to 'power' in its widest sense. In terms of gender, 'where' women are positions them in a range of contexts and in different and changing relations to other women and to other places (Massey, 1996).

Turning to our second point, feminists have also shown the diverse ways in which femininities are constituted through specific sites, spaces and networks. The work-place, for example, is one particular space in which women must negotiate their femininity. Two contrasting examples of the ways in which women negotiate femininity at work can be seen in Pringle's (1989) study of secretarial workers, and McDowell and Court's work on women employed in financial institutions in the City of London (McDowell and Court, 1994a). Feminist geographers have also started to look at the home as a site for the negotiation of femininity, taking into account the range of experiences different women may have in this space. Home can be a site of refuge from racism for black women, or a site that must be reworked to challenge dominant heteropatriarchy for lesbians (Valentine, 1993; Johnston and Valentine, 1995). Similarly, feminists have also considered the ways in which femininities are constituted through different local social networks. For example, local neighbourhood networks, or 'communities', can provide specific spaces within which femininities are negotiated – often through ideas about work and/or motherhood (Bell and Ribbens, 1994; Pratt and Hanson, 1995; Dyck, 1996; Holloway, 1998b). These networks provide a space in which local moral geographies are constructed and reconstructed, shaping and reshaping local understandings of 'appropriate' femininities.

A focus on the spaces of work, home and neighbourhood starts to illustrate our third point – that our understandings of different spaces are constructed through discourses about gender. The previous section on the historical geographies of femininity showed how hegemonic definitions of femininity in Britain, at different points in time, have reinforced the notion that the home is a feminine sphere. This association is naturalised, with the home considered the site of motherhood and domesticity. Similarly, feminist geographers have considered the different ways in which public space can become encoded by gender. For example, spaces such as the streets (Valentine, 1989) or parks (Burgess, 1998) are considered unsafe for women and particular parts of the city can often be seen as masculine or feminine (Bondi, 1998). It is not simply that femininities are constructed and contested through different spaces, but also that our concepts of spaces are themselves actively produced through notions of gender.

In summary, this three-fold approach to the geographies of femininity not only emphasises that there are 'geographical variations in what it means to be a man or a woman, but that our very notions of masculinity and femininity (in all their subtle variations) are actively constituted though distinctions of space and place, public and private, visible and invisible' (Jackson, 1993: 222). Consequently, in this book, we want to emphasise that place and space play a crucial, active and constitutive role in the ways in which femininities are (re)formulated. It is this intertwining between geographies and femininities that is our key focus of enquiry.

Outline of the book

This intertwining of geographies and femininities is first opened up in Chapter 2 when we consider two sets of literature – one focusing on globalisation, the other considering identity formation – that are too often considered in isolation. We concentrate first on globalisation and highlight key processes of economic, cultural and political change. We then move on to look at work that theorises the production and contestation of identities as a multiple and fractured process. We pinpoint the nexus of the globalisation and identities literatures, which we argue have much to offer to our understandings of the intertwining of geographies and femininities, through concrete examples focusing on the gendered impacts of globalisation and the (re)production of gendered identities. This material provides much of the conceptual framework for the four case studies presented later in the book.

These case studies, two based in Britain, one in eastern Germany and one in Peru, are first introduced in Chapter 3 where we consider the ways in which each of us has translated our theoretical interest in geographies and genders into empirical, case-study research. We begin by outlining briefly the feminist debates about methodology in geography and highlight how we worked through issues of power and positionality in 'the field'. We then move debate forward with a consideration of the ways in which fractured and fracturing understandings of gender can be worked with in empirical studies, and by exploring how our three-fold typology of geographies and femininities (see above) can be mobilised in practice.

The central section of the book contains four individually researched and authored, but collectively edited chapters focusing on each of our case studies. In Chapter 4 Nina Laurie considers women's involvement in state emergency employment in Peru, and looks in particular at the multiple representations women make of themselves to each other and their husbands/partners. Sarah Holloway focuses in Chapter 5 on two arenas where the reproduction of ideas about good mothering can be most clearly seen in Britain – namely around issues of motherhood and paid employment, and in attempts to secure a pre-school education for young children. In Chapter 6 Fiona Smith examines the politics of gender identities in the radical changes of post-communist 'transition' and considers specifically how women negotiate political citizenship in neighbourhood action groups in eastern Germany after reunification. Finally, in Chapter 7, Claire Dwyer analyses the importance of new Muslim identities and other social differences in young Muslim women's negotiation of identities both at home and at school in suburban London.

We believe that this theoretically informed empirical work, and in particular the comparative approach that the use of our four case studies allows us to pursue, has much to offer our understanding of the geographies of new femininities. This theme is explored in Chapter 8 where we begin by considering the ways in which geographies and new femininities are mutually constituting, and give particular emphasis to the importance of place, the spaces of paid work, home and neighbourhood, as well as gendered discourses about these spaces. This analysis highlights important aspects of the intertwining of geographies and femininities in and across space. In particular, the chapter moves on to discuss a range of ways in which geographical thinking about processes of change may need to be reconfigured. The chapter ends with an exploration of the significance of geographical perspectives for interrogating understandings of how identities change.

In the conclusion to the book we return to some of the themes first raised in this chapter. We reiterate our insistence on the importance of studying a diversity of new femininities rather than simply those hyped within the (British) media. We then reconsider the importance of the historical geographies of femininities in Britain, Germany and Latin America outlined above in the light of the four case studies. We suggest that these elucidate the meaning of the term *new* femininities and that they continue to be important in the everyday production of new femininities. In considering the reworking of femininities we not only celebrate the ways in which some women are able to renegotiate their gendered identities, but also highlight instances where this is not the case and where change may be highly problematic. Indeed, we are careful to make clear that the emergence of new femininities is not necessarily a progressive phenomenon, and moreover that different women will define what is 'progressive' in different ways. We end the book with a summary of our argument and the suggestion that an appreciation of geographies is central in understanding new femininities, just as new femininities should be a central feature of feminist geographers' future research agendas.

CHAPTER TWO

Changing worlds?
Changing femininities?

Introduction

This chapter explores two sets of literature within geography and the social sciences which, we argue, when brought together can further our understanding of the connections between geographical processes and the construction of new femininities. First, we consider the body of work on current economic, cultural and political processes sometimes collectively described as 'globalisation'. This work has arguably 'become a defining drama and preoccupation of English-speaking social science in the nineties' (Ó Tuathail, 1998: 85), and is certainly of crucial importance in studies of changing geographies at local, regional, national and international levels. Rather than provide an overall survey of this literature, we draw out those aspects of relevance to this particular book grouped around themes of economics, culture and politics. In the following section we then move on to consider the development in recent years of the ways in which social scientists, and particularly feminist geographers, work with and theorize the production and contestation of identities.

While these two sets of work are often considered in isolation from one another, they are increasingly being brought together to provide new ways of teaching about and researching the world (see Figure 2.1). This trend reflects the influence of feminist analyses, which have long been critical of the tendency to divide the 'social' from the 'economic' (WGSG, 1984), as well as other factors such as the recent 'cultural turn' in economic geography (Lee and Wills, 1997). In the final section we draw out points of intersection between the broadly defined 'globalisation' and 'identities' literatures through an analysis of how different processes may constrain women's lives, open up spaces for new opportunities, or be contested by the actions of women in a range of spaces and places. In particular we are interested in the gendered implications of globalisation and the (re)production of gendered identities, a focus that includes an interest in the ways in which gender identities are simultaneously racialised, classed, sexualised, and so on, through different geographical processes.

WHAT IN THE WORLD

Stahler

Figure 2.1 New ways of teaching about divisions in the world. © United Feature Syndicate, Inc. Reproduced by permission

Globalising economies, cultures and politics

'Globalisation' is perhaps the most common description of current processes of global and local change in geography and the social sciences, yet considerable debate exists about exactly what constitutes globalisation and what its effects might be. On the one hand writers such as Hirst and Thompson (1996) argue that 'globalisation' is merely a powerful 'myth' which, when examined on the ground, cannot be proved. On the other hand some authors such as Luke (1997) offer specific examples of globalisation in progress, pointing to the changing circulation of capital, division of labour and integration of markets in a range of situations. In the context of this polarisation of globalisation debates many authors are finding a middle ground, arguing that there are both elements of 'truth' and of overstatement in the analysis of the processes generally known as 'globalisation' (Dicken *et al.*, 1997). The purpose of this section is not to engage specifically with this debate, to prove or disprove the existence of globalisation, but rather to identify some key global and local processes of economic, cultural and political change in order to frame the later discussions about emerging femininities which form the core of this book. Specifically, we examine the processes associated with the economics of globalisation and the rise of neo-liberalism. We address questions about the rise of increasingly global cultures and globally mediated knowledges and related issues of the changing power of the nation-state and the emergence of new forms of political action.

Globalising economies?

According to Castells (1997), globalisation is currently characterised by a network society, produced by the coming together of two processes: a revolution in information technology and the restructuring of capitalism. Companies seek new strategies for improved profits through measures that, for example, utilise an international division of labour and production. Such measures rely upon technological changes while simultaneously driving the search for communication improvements. Globalisation, in part then, refers to global flows of capital and ideas at new speeds, over longer distances, facilitated by advances in electronic communications and by the altered international regulation and operation of financial markets (Leyshon and Thrift, 1996). Processes of time–space compression and time–space convergence profoundly alter (though not as profoundly as is sometimes suggested) the connections between economy, place and space (Harvey, 1989; Leyshon, 1995).

A whole range of studies illustrates the importance of changing technologies in enabling and structuring flows of capital and labour, and restructuring the relations of employment and control between different locations in unequal and uneven ways (Harvey, 1989; Allen and Hamnett, 1995). Pearson and Mitter (1993), for example, show how the rise of offshore data processing, made possible through technological change, may simply enable business to exploit wage differentials between countries and regions, moving low-grade tasks to less unionised and lower paid locations such as India or Jamaica. In turn much of this labour force is made up of women who constitute in many countries a source of 'green' labour which is plentiful, cheap and supposedly biddable. One key feature of globalised labour markets is therefore the global increase in women's involvement in paid work.

The extent to which production processes have become mobile and global markets increasingly interconnected, so that a country's stock market collapse can be brought about by trading decisions on the other side of the world and collapse can 'spread' between national and regional economies to threaten global crisis, all suggest that space may indeed have been 'annihilated' (Leyshon, 1995). However, in other senses space and place matter in very particular ways to the operation of this apparently global system. Thrift and Leyshon's (1994) study of the City of London illustrates both the deterritorialised nature of the global flows of the financial markets (their detachment from notions of bounded territories 'on the ground', such as the nation-state or the region), and the specific, localised processes of interpersonal linkages in trading floors and the urban spaces of the City of London itself on which the processes of global capital circulation rely. The de-linking of capital from space also depends on capital being anchored (perhaps temporarily and selectively) in specific local places, which both shape and are shaped by social, personal and institutional processes. A variety of social interactions and power relations are therefore embodied in institutions which mediate global flows.

At the continental level circulation is mediated through the context of new alignments of capital as geographies of commerce are reconfigured around new trading blocs (Brook, 1995) such as the European Union (EU), the North American Free Trade Association (NAFTA) covering the USA, Canada and Mexico, or MERCOSUR in Southern Latin America and the Andes, as well as new associations in Africa and Asia. In many cases trading relations take on new political meanings which have both a transnational and sub-national focus. In the case of the EU, for example, legislative changes and monetary union affect state policy while a 'Europe of the Regions' may offer more scope for the self-determination of regional and collective identities (Amin and Tommaney, 1995). As Marchard (1994) writing on NAFTA points out, these emerging trading blocs also establish new inclusions and exclusions. She highlights in particular the role of female '*maquiladora*' labour in NAFTA, where women are drawn into working in export-oriented production but do so on very poor terms. Studies on EU policies have pointed to a range of explicit and implicit exclusions around race, class, gender and sexuality involved in the structures of European integration (Kofman, 1995; Vaiou, 1995; Valentine, 1996).

No matter which interpretation of globalisation is adopted, we want to emphasise here that in recent decades globalising economic practices have become increasingly associated with a particular ideological agenda for globalisation, namely neo-liberalism. Neo-liberal paradigms promote a series of economically driven adjustments which include privatisation, switching policies (a shift towards production for export and away from production for domestic consumption) and cut-backs in the public sector. Together these processes constitute a 'rolling back of the state' designed to encourage private capital and non-governmental organisations (NGOs) to take the lead in socio-economic management. Neo-liberal agendas have been common in many western countries where governments have engaged in widespread programmes of privatising state-owned companies and utilities, and also in introducing 'quasi-markets' in public service provision from health-care to education (Marvin and Graham, 1994; Peck and Tickell, 1994; Martin, 1997). In recent years, in many countries of 'The South' and the former Eastern Bloc, neo-liberal agendas have been introduced particularly, though not exclusively, through structural adjustment packages (SAPs) implemented at the behest of international lending agencies keen to recoup the losses that occurred during the so-called 'Third World debt crisis' of the 1980s and to structure the newly emerging capitalist economies in Eastern Europe and the former Soviet Union (EEFSU). In the late 1990s many Asian and Latin American countries also face structural readjustment following turmoil in currency and stock markets (Higgins, 1998). As we discuss later, women often bear the brunt of these adjustments when their access to land and free public services decline, and their domestic work is compounded by employment and greater involvement in community management during economic crises.

Amin (1997) suggests that globalisation and neo-liberalism are often uncritically conflated and are frequently seen as synonymous in popular and academic writing. The relationship between the two, however, can be usefully unpacked. In many parts of the world globalisation processes play a significant role in the spread of neo-liberalism. Laurie and Marvin indicate the specific ways in which globalisation promotes neo-liberalism:

> Global networks are facilitating the dissemination of information and ideas about neo-liberal strategies. International donor organisations are sponsoring social programmes designed to cope with the supposedly short-term social costs of 'shock' economics in order to facilitate what the World Bank calls 'adjustment with a human face'. And increasingly, global companies are buying into newly privatised industries . . . globalisation also provides the context for the international policing of neo-liberalism [as] global institutions such as the IMF, the Inter-American Development Bank and the World Bank promote 'global answers' to what they see as 'global problems'. These organisations espouse an international free market on one hand and yet, on the other hand, impose sanctions and withhold funding in individual countries if adjustment measures are not followed to the letter. The policing of the neo-liberal package is therefore global.
> (Laurie and Marvin, forthcoming)

Barton (1997: 77), extends this analysis of policing into Latin America by suggesting that, into the next millennium, US engagement with Central and South America will be shaped by 'global neo-liberal political economy and the processes of globalisation that characterise it'.

A key moment in the development of the processes of globalisation is found with the collapse of communism in the late 1980s and early 1990s. The subsequent exposure of the economies of the states involved to global competition and the turn to market-economic and democratic systems 'has resulted in a deep and thoroughgoing modernisation of the structures and institutions of government, economy and social life' (Smith and Pickles, 1998: 5). Possibilities for a 'third way' of reformed communism with a 'human face' were largely swept aside in the optimism of early transitional processes, and widespread anti-communist feeling. Also important at this time was the rise of a general public perception of the 'West' as offering wealth, freedom and opportunities (Altvater, 1993). This was encouraged by western propaganda and the specific institutional politics of the 'machine of transition' (Smith and Pickles, 1998: 4) provided by bodies such as the World Bank and the IMF. These and other organisations, including western governments, provided international assistance which involved transfers of 'know-how' from West to East. In part the electoral politics of post-communist states have been shaped by the questions of whether such changes are necessary or inevitable, and whether the cost to the populations in terms of unemployment and cuts in social expenditure, for example, are outweighed by the benefits of change.

As suggested above, the politics of 'transition' have been dominated by the agendas of neo-liberalism. However, it must be noted that it would be wrong to view these policies as ones being advocated only by particular institutions or actors. These principles are regularly advocated by EEFSU governments and sections of their national electorates alongside western donor agencies. Overall, the general trend of development in the post-communist states constitutes an example of widespread change shaped by discourses and related institutions which conflate neo-liberalism, globalisation, modernisation and development, something that is not restricted to EEFSU (Sidaway and Power, 1998) or to post-communist states. Nevertheless the experiences of post-communist restructuring do entail quite specific ideological assumptions about the 'end' of communism. In the hegemonic neo-liberal discourse, communism is relegated to an outmoded historical fact, rather than a possible ideological alternative to capitalism, which is now celebrated as victorious. In this situation the massive difficulties faced by the states, economies and the populations of post-communist countries are typically positioned as inevitable problems associated with the prize of achieving western-style capitalism, with blame placed on the continued legacies of the failures of the previous system or the inability of the population and of institutions to internalise the appropriate culture required in the new system. Unemployment, for example, becomes defined in the logic of the free market not as a failing of the market but as the inability of individuals to adopt the requisite entrepreneurial character (Bridger et al., 1996).

The understandings of space in such interpretations of post-communist 'transition' are quite particular. At the global geopolitical scale, the 'end of communism' and the ending of Cold War divisions are seen to mark the 'triumph' of capitalism and its now-expanded reach to cover most of the globe. The processes of privatisation, democratisation and liberalisation are universalised, posited as both inevitable and desirable for all places, and the significance of the particular and specific experiences of pre-communist, communist and post-communist development of different countries, regions and places are by and large ignored. Most distressingly, perhaps, this assumption that the destination of 'transition' is known and is automatically progressive is questioned fundamentally by the ferocious nationalism and ethnic conflicts of many post-communist states. Those displaced by the creation of new borders are all too aware of the specific regional characteristics of post-communism (Pilkington, 1997). This attention to the difference that place makes in these processes has been taken up in a range of studies that examine the particular 'pathways' of post-communism. Work in this field stresses that the combinations of past and current developments in specific localities do not simply produce 'aberrant' forms of capitalist development, but rather that the very varied dynamics of local and regional processes produce a variety of trajectories of 'transition from state socialism' (Grabher and Stark, 1997; Smith and Swain, 1997; Dunford, 1998). In fact, the variety of pathways of change suggests that far from the dynamic assumed by the 'universalism and triumphalism of transition theory', 'transition is not a one-way process of change from one hegemonic system to another [but rather] constitutes a complex reworking of old social relations in the light of [. . .] attempts to construct a form of capitalism on and with the ruins of the communist system' (Smith and Pickles, 1998: 2). In these conditions, then, the effects on women's involvement in the labour market and in social and familial spheres, and the intersections of gender identities with other changing identities and their politics, are complex, involving renegotiation, rejection and continuation of past and present positions.

Globalising cultures?

Globalisation may be understood not only as an economic phenomenon but also as a cultural phenomenon. Many studies point to the connections between increasingly global cultures and globally mediated knowledges. We need only to think about the widespread circulation of the images of new femininities discussed in Chapter 1 to illustrate what Morley and Robins (1995) identify as the increasing importance of communications technologies in circulating globalised popular cultures. They argue these are, in turn, reworked and appropriated in particular contexts, and possibly even recirculated as globalisation generates the (imperfect) transfer of knowledge and ideas through the local–global nexus created by time–space distanciation. This has resulted in debates around the question of whether globalisation promotes the homogenisation of culture or creates an opportunity for the construction of new forms of heterogeneous or 'hybrid' cultures (Hall, 1995; Morley and Robbins, 1995). It is also suggested that increasing levels of interconnectedness may provide a new space of identification through an experience of encounter and solidarity (Morley and Robins, 1995). The ability to connect beyond the local may be a liberation or a revelation for some and may indeed offer possibilities for reworking identities, as we consider in relation to gendered identities later in the chapter. However, most recognise such technologies never escape the uneven and unequal social relations from which they arise. The overwhelming dominance of male producers and consumers of IT, for example, or the global, regional and local inequalities in access to technologies may only serve to entrench existing inequalities (Kitchin, 1998).

The globalisation of culture is not simply technologically driven, however. Other powerful forces of globalisation are migration and diaspora. It has been argued that we now live in 'an age of migration' (Castles and Miller, 1993) in which global migration movements are accelerating at an unprecedented rate. This increase in the numbers of migrants is also paralleled by the *globalisation* of migration – more and more countries are becoming incorporated within wider systems of population movement, of both skilled and unskilled workers, legal and illegal migrants. Such movement arises out of processes of uneven development, operates through pathways of colonial and neo-colonial linkages, and in turn shapes the very nature of economic and cultural processes of globalisation as it allows the further mobilisation of labour for the profits of capital. Yet, it also shapes the sites and linkages through which globalisation proceeds, in the shape of regional labour markets, for example, or in the development of global cities (Sassen, 1991; Beaverstock, 1996a, 1996b; Pettman, 1996; Li et al., 1998).

Global population movements generate transnational linkages and 'stretched out' social relations (Massey and Jess, 1995) producing 'diasporas', peoples who share cultural affiliations and identities but who are dispersed across geographical boundaries (Hall, 1995). Such populations may result in diasporic or 'hybrid' cultures as cultures are 'transmitted' and transformed within new locations and contexts, creating new forms; developing them in new ways;

or giving new appreciations of difference and diversity. Such processes of encounter and change may be celebrated as transculturation (Hall, 1995) whereby bounded notions of cultural difference are challenged. The questions that such processes have raised for issues of identity are discussed further in the final section.

The globalising demand for the consumption of other cultures, however, does not only arise on the basis of the movement and mixing of people but also simply from opportunities sought by capital for further profit gained by commodifying difference as those with money are offered the chance to consume ethnic difference. May (1996) argues that some members of the new cultural class in Britain sustain their 'class' position through consumption rather than employment. For example, 'foodies' cook and eat exotic food rather than mainstream produce available at the local supermarket. While many of these people pride themselves on their liberal credentials, their attitudes to food are based on a host of racist stereotypes which are sometimes carried over into their attitudes to other residents in their local area. This commodification is increasingly common and is based on making cultural difference exotic. hooks (1992: 21) suggests that 'the exotic' may be seen by these consumers primarily as 'spice, seasoning that can liven up the dull dish that is mainstream white culture'. Consequently the demand to consume other cultures by people in privileged positions may serve to re-enforce existing global power relations and prejudice.

Responses to these processes of change vary considerably. While in some places globalising cultural processes may be welcomed and may be genuinely transformative, in other cases the response to globalisation of culture may be a reassertion of local identifications and a stress upon more bounded notions of belonging. This defending of 'traditional' or 'fixed' cultural values may have important gendered implications. For example, women are frequently seen as bearers of tradition and culture, and representations of femininity often play an important role in the maintenance of ideals of nationhood or cultural belonging (Johnson, 1995; Sharp, 1996; Yuval Davis, 1997). These responses are not simply cultural but constitute in turn part of a changing political response to, and engagement with, processes of globalisation.

Globalising politics?

New forms of political identification beyond the nation-state are being made possible through the growth of transnational linkages encouraged by both increased communications and the development of diaspora populations across the globe. The growth of globalised support networks for Hindu nationalism in India (Rai, 1995) or diaspora Arab politics are both good examples. Such forms of political affiliation and identification may work to challenge the authority invested in the nation-state. A further example is provided by Radcliffe's (1998) analysis of the 1995 Peru/Ecuador border war. She shows how indigenous groups, whose territories cross the borders of the two nation-states, and women's groups in both countries condemned the war using their international networks. In so doing they expressed transnational solidarity and contested the border. They also appropriated the new technologies associated with globalisation (faxes and e-mail connections) in order to communicate their resistance widely.

While such transgression of borders may enable some to pursue particular politics of resistance, the reverse is also true. Enforced crossing of borders and displacement in the past 20 years has produced unprecedented numbers of refugees and asylum seekers (King, 1995), the majority of whom are women (Martin, 1992; Agger, 1994). It is ironic then that as global population movements have increased, so has the drawing of borders or boundaries of exclusion which prevent the gaining of citizenship in other countries. This is clearly illustrated by the 'fortress Europe' immigration policies of the EU and its member states which have coincided with increased freedom of movement within the internal boundaries of Europe. Thus, refugee diaspora populations may find themselves rendered powerless and stateless while other groups of people move more freely as they become 'citizens of Europe'.

In addition to these transnational politics, we have already seen that economic processes of globalisation and the global policing of neo-liberalism raise important questions about the changing shape of political power. These processes have led some to claim that the power and importance of the nation-state is withering (Ohmae, 1990; Horsman and Marshall, 1994) as individual states are now less able to determine their own policies (though whether this was ever entirely true, rather than a powerful fiction, is difficult to evaluate). Spectacular examples such as the crash of many South-East Asian economies in 1997/98 and the knock-on effects of these on financial markets and job losses across the world certainly indicate the 'deterritorialisation of power', as individual countries are less able to control their own economic fate in the face of fluctuating global financial markets. The rise of regional economic and political groupings built around trade and/or political and military interests, such as the EU or NAFTA (mentioned above), can equally be seen as a shift of power from nation-states to alternative institutions, which are often designed to cope with the economic vagaries of globalisation by creating favoured trading areas at a supranational scale (Ohmae, 1995).

However, others argue that the 'death' of the nation-state is 'exaggerated' (Anderson, 1995), and dispute any notion that 'geography is history' (as an advert for British Telecom recently proclaimed). Instead power is 'reterritorialised' in selective ways that reassemble new geographies of power (Ó Tuathail and Luke, 1994). With the end of the Cold War, President Bush did not declare the end of a global order, but a 'New World Order' under primarily American dominance. The collapse of Soviet influence allowed the emergence of nationalist and regionalist campaigns resulting in the establishment of a range of nation-states. These contradictory tendencies have led some commentators to argue that nation-states themselves, though not necessarily withering, now mobilise power in different ways. Ruggie (1993) supports a reconsideration of sovereignty, suggesting it is increasingly 'unbundled' from attachment to a bounded national territory, residing in a 'space-of-flows' (capital, ecological ideologies, information technology and communications) as well as a 'space-of-places' (state territories, economic wealth in particular locations, the attachment to national ideas). This recognises that control over the means of sanction, and in some cases repression, remains in many cases firmly in the power of the state. For people subject to such control, theories of the deterritorialisation of power surely seem fanciful. Castells (1997) suggests that today's states are in fact caught between contradictory processes of globalisation. He argues that while states seem to be losing power, they are not actually losing influence:

> The fading away of the nation-state is a fallacy. However, in the 1990s, nation-states have been transformed from sovereign subjects into strategic actors, playing their interests, and the interests of those they are supposed to represent, on a global system of interaction, in a condition of systematically shared sovereignty. They marshal considerable influence, but they barely hold power by themselves, in isolation from supranational macro-forces and subnational micro-processes. (Castells, 1997: 307)

In terms of their own populations, states retain considerable importance (though not perhaps absolute independence) in their policy decisions on a vast range of domestic issues from the scale and form of service provision, to tax and welfare regimes and the shape of the political system itself, including the definitions and practices of citizenship. However, the pervasiveness of neo-liberal restructuring across states (whether so-called First, Second or Third World states) has meant the rise of forms of 'governance' that are not strictly within formal elected political spheres. This includes the appointment of semi-private development agencies for regional economic planning, or the transfer of public resources into forms of private or at least non-state provision (Kearns, 1995; Laurie and Marvin, forthcoming).

Just as several of the examples in earlier parts of this section have argued, it is a mistake to assume that power might be shifting in one direction (namely to the increasingly globalised power of capital). Globalisation is simultaneously global and local, and while the role of the state may be changing in the context of processes of globalisation, so might other forms of less 'formal politics'. Many argue that, for example, the rise of 'new social movements' (NSMs) is directly influenced by the power of globalised spaces and networks (Escobar and Alvarez, 1992; Castells, 1997). These social movements represent 'new forms of doing politics' because as primarily 'grass-roots' organisations they are assumed to be more representative and democratic. Their informal structures contrast with the clearly defined hierarchies and power relations of formal politics. They are based on collective action, often give a high profile to women and are seen to coalesce around issues of identity and interests of everyday concern. In the 1980s many authors assumed that NSMs were displacing more traditional forms of social protests such as class-based struggles. More recently, however, the continuities between these movements and more established traditions of resistance, such as urban class struggle, have been emphasised (Canel, 1997; Alvarez *et al.*, 1998). While NSMs mobilise for a variety of reasons and attach different meanings to their activities, they are all characterised by 'their continuous struggles against the dominant projects of nation building, development, and repression' (Alvarez *et al.*, 1998: 6). Many such social movements have the sort of transnational linkages discussed above and, in particular, have been seen as key forces in resisting neo-liberal adjustment. A particularly good example of this is the internationalisation of the women's movement in recent years. Women's movements in countries of 'The South' and 'The North' have been networked for nearly two decades through gender and development initiatives. While international forums such as the UN Women's Conferences in Nairobi, 1985, and Beijing, 1995 have their critics, such international encounters and advocacy networks nevertheless represent a gender-based transnational identity politics (Sikkink and Keck, 1998).

In this section we have highlighted the different processes that might be linked together within a definition of globalisation. We have emphasised, in particular, economic changes such as the rise of the 'network society' and the restructuring of capitalism (and the concomitant forces that have led to the dominance of neo-liberalism) and the ways in which globalisation encompasses cultural changes through both technological influences and global population movements. We have raised questions about the extent to which globalisation has been accompanied by political changes, particularly in relation to nation-states and their policies but also in relation to transnational and informal political action. Our primary interest in outlining these processes is to establish a framework in which we ask how these interconnected global and local processes of change are of significance in shaping the changes in women's lives in different places around the world. To what extent might they either constrain women's lives or open up spaces for new opportunities for women? How are these processes contested by the actions of women in a range of places and spaces? In the final section of the chapter we focus more specifically on such gendered geographies, but before that, in the next section, we address some of the questions raised about new forms of identification in the context of social sciences literature on identity and the politics of difference and ask how feminist geography in particular has engaged with issues of identity.

Fractured and fracturing identities

Debates about identity and the cultural politics of difference have also been a prominent concern of the English-speaking social sciences since the early 1980s. We begin this section with an account of this growth of interest in identity and difference, and the widespread rejection of essentialist and universalistic assumptions about identity that this has involved. This leads us to an exploration of the importance of identity politics and its relation to more 'traditional' class-based forms of political organisation. In this analysis, it becomes clear that the articulation of multiple forms of social difference raises important questions for both 'Leftist' politics and the very NSMs that are often based upon specific identities. We then turn from this general analysis of identity to a more explicit consideration of the ways in which feminist geographers have worked with difference, and in particular how our understandings of gender have become increasingly fractured over time. The section ends with a consideration of post-structuralist work, and in particular that which has drawn our attention to the importance of our ideas about the body and bodily performances.

Identity and the cultural politics of difference

The literature on identity and difference, and on identity politics (or the cultural politics of difference as it is sometimes called), has grown rapidly since the early 1980s. Much of this debate has been concerned with the production of gendered, sexualised and racialised identities, though more recently other axes of difference, such as disability or age, have begun to receive greater attention. A common strategy, in what is admittedly a rather disparate set of writings, has been to dispute essentialist and universalistic assumptions about identity, that is that identities are in some way biological or natural, and argue instead that identities are created through processes of social construction. Thus, as we highlighted in Chapter 1, feminists have argued that the differences between femininity and masculinity are not natural, but rather that girls and boys, women and men, are constantly being (re)taught how to behave in appropriately feminine or masculine ways. Individuals may absorb or contest such hegemonic definitions of appropriate gender behaviour in constructing their own gendered identities, but they always do so in a context where some decisions are easier to make than others.

An important contribution to the retheorisation of identities is the work of Stuart Hall and others (Hall, 1990, 1992b; Gilroy, 1993) in redefining ethnicity not as an essentialist identity – a fixed attribute conferred by birth or parental background – but as a dynamic concept. Ethnicity, Hall (1992b) argues, is shaped by particular historical and cultural discourses, but is also a social and political construction, and hence is always in process, always being made and remade within particular social and political contexts. As Hall argues, 'we all speak from a particular place, out of a particular history, out of a particular experience, a particular culture, without being contained by that position' (Hall, 1992b: 258). Similar issues are also important in relation to other identities, but they are not always mobilised in the same way. For example, in the case of lesbian and gay identities the essentialism versus social constructionism debate is somewhat less clear-cut. Some activists and academics argue that homosexuality does indeed have a biological basis and use this argument to political effect (if it is natural it cannot be wrong). Others argue that this position is inherently one of political weakness, and argue instead that sexuality, like gender and ethnicity, is a social construct, shaped in different ways in different times and places (Bell and Valentine, 1995).

The identities with which these debates are concerned are at once personal and political. Your identity – who you are, how you feel about yourself, how others respond to you – is about you as a person. However, because identities are socially constructed they are not simply personal issues, they also place you (and 'us') within social structures that confer power and wealth on some, and disadvantage on others. Rutherford (1990: 12) notes that much of the work

on identity has been 'formulated from marginal positions'. This perhaps is unsurprising as the very marginality of some groups can make their identity more
noticeable. The performance of hegemonic forms of masculinity, heterosexuality or whiteness (in masculinist, heterosexist and white-racist societies) often
goes unnoticed, whereas people who perform, for example, non-hegemonic
femininities, lesbian and gay identities or minority ethnicities are labelled as
different, deviant or challenging to social stability. In order to challenge the discrimination or disadvantage associated with minority identities, critical academic
work (research and teaching) has focused on both validating these identities and
critiquing the processes through which disadvantage is reinforced.

Interest in the political importance of identity is by no means confined to
academia: ' "Identity politics" has been current for some time as contextual
shorthand for movements organising around sexuality, gender and ethnicity
and working to translate "the personal is the political" into everyday practice'
(Brunt, 1990: 151). These forms of political organisation are sometimes refered
to as new social movements (which we have defined in general terms earlier
in the chapter). Their newness refers to the sense that they are organised around
issues other than class, and (as mentioned earlier) to their often informal rather
than formal organisation. The links between identity politics and class politics,
between NSMs and more 'traditional' forms of political organisation, have often
been difficult. For example, in 1980s and 1990s Britain and the USA the New
Right mobilised exclusionary discourses – playing up their desire to 'protect'
the country from further waves of immigration or from the proliferation of
'non-traditional' households, especially those headed by lone parents – to further
their own political agenda, gaining popularity by distancing themselves from
these socially and/or economically marginalised groups. The Left, broadly defined,
who might be expected to embrace these 'minority' interests has also had an
ambivalent relationship with identity politics. For some on the Left identity
politics represents a bourgeois distraction from true political action, that is action
based around class politics (Brunt, 1990; Studzinski, 1994). Perhaps more
fundamental, though, are the difficulties the Left experiences when trying to
embrace these differences. Again drawing on a British example, Rutherford
(1990: 19) argues that where 'the Communist Party's broad democratic alliance
failed and municipal socialism's engagement with the new social movements
went wrong was in the assumption that these new political identities were fixed
categories'. In other words, the Left tried to encompass an appreciation of these
other differences alongside its interest in class, but failed to recognise that these
differences are not fixed but articulate with one another.

This articulation of the multiple forms of social difference has been an important theme in the study of identity and identity politics. Initially, there was a
tendency to think of difference in additive terms; thus there was talk of the
'triple oppression', that is being black, working class and a woman. Over time,
however, there has been a growing appreciation that these differences are mutually constituted. As we outlined in Chapter 1, for example, an individual is not
a woman and white in any simple sense. She does not have one set of experiences that stem from her gender to which is added one stemming from her 'race'.

Rather these subject positions are shaped together producing a racialised femininity, or a feminised ethnicity (Brand, 1990; Ware, 1992). This mutual constitution of the axes of difference complicates calls to political action: there is no simple social group such as 'the workers', to whom appeals for social action can be made, as the category 'worker' is fractured by a variety of other social differences (which are themselves fractured by class).

Consequently, there has been continuing ambivalence from many Leftist academics to identity politics. Harvey (1993: 61–62) shows his frustration with (what he calls radical post-structuralist) debates about difference, arguing that they have led to the loss of 'political and ethical solidarities and similarities across difference' and proposes instead a re-engagement with the concept of political economy. His argument appears to be that a politics based on class interests is preferable to no political action at all: 'pursuit of working-class politics might protect, rather than oppress and marginalise, interests based on gender and race even if that working-class politics regrettably makes no explicit acknowledgement of the importance of race and gender' (Harvey, 1993: 59). In addition, the question of how to work with difference within minority groups themselves is also important. As Rutherford (1990) points out

> the paradox is that when the margin resists and discovers its own words, it not only decentres the dominant discourses and identities that have suppressed it, but also transforms its own meaning. Just as it invades the centre with its own differences, so it too is opened up to its internal differences. (Rutherford 1990: 23–24)

As the next section discusses, one of the most important (though by no means only) ways in which these debates about identity and identity politics have been brought into geography is through the work of feminist geographers.

Gender, identity and feminist geography

Feminist geographers have shown how ideas about gender are shaped in different ways in different times and places, and have worked to highlight and challenge the negative consequences of this in the everyday lives of women (and men). In doing so they have also demonstrated the importance of spatiality in identity construction, highlighting how gendered identities are constructed in and through particular sites, spaces and networks within particular places (see Chapter 1). The emergence of this work in the 1970s, and then on a much wider scale in the 1980s, presented a challenge to both mainstream and radical geography of the time which had marginalised the study of women and gender, often inadvertently. In Monk and Hanson's (1982: 12) words: 'Most academic geographers have been men, and they have structured research problems according to their values, their concerns, and their goals, all of which reflect their experience'. Like many elsewhere, feminists in geography responded to this challenge by focusing on the singular category 'woman'. This was a useful strategy at this point in time; however, more recently feminist geographers have been working out ways to deal with diversity and difference (WGSG, 1997).

In fact, debate about the intersection of class and gender arose relatively quickly within feminist geography, though the debate was framed more in terms of social structures than of subjectivity. A number of feminist geographers had their roots in radical geography, and brought to the sub-discipline a socialist–feminist analysis that examined the importance of gender differentiation to the capitalist system (McDowell, 1983; Mackenzie, 1984). Others worked within a radical feminist framework, and employed a realist analysis to argue that capitalism and patriarchy (the system of social relations which works to the general advantage of men not women) were separate but interacting systems (Foord and Gregson, 1986). The debates between these two groups of feminists were most notably played out on the pages of the journal *Antipode* in the mid-1980s (McDowell, 1986; Gier and Walton, 1987; Gregson and Foord, 1987; Johnson, 1987; Knopp and Lauria, 1987).

Despite its focus on the links between, or unity of, capitalism and patriarchy, this debate also raised for the first time the heterosexism inherent in feminist thinking in (British) geography. In their original article Foord and Gregson (1986) had argued that heterosexuality is a necessary relation of gender relations, but in their responses Knopp and Lauria (1987) and Johnson (1987) both argue that this cannot be so as heterosexuality is not transhistorical and transspatial; rather it is socially constructed in specific times and specific places. The specifics of this exchange are perhaps less important than the fact that this was the first time the question of heterosexism had been raised in connection with the way (British) feminist geographers were theorising gender. Janet Townsend, in a reflective account of her experiences as a feminist geographer for the WGSG (1997) book, describes how British feminist geography in the 1980s, concerned as it was with patriarchal gender relations, failed to consider other social differences and thus shaped its research agenda around issues of interest to heterosexual feminists:

> Dutch geographers opened my eyes to a central feature of geography. They invited members of the group to a week's meeting in Amsterdam, and kept asking us 'Why do you always write and talk in terms of the nuclear family?' We felt insulted, the nuclear family being an orthodox enemy, until we realised that the real criticism was of heterosexism. (WGSG, 1997: 54)

In effect many feminist geographers, in a similar way to the male establishment before them, were shaping their research agenda around their interests which reflected their experiences as heterosexual women.

Their research agenda also reflected the fact that most were white women. For example, white feminists, and white feminist geographers among them, have been very critical of the western conception of the home as a haven from the pressures of the outside world, arguing that the home can be a site of patriarchal oppression and extreme loneliness for women. In response to this universalising discourse on the home, black feminists have pointed out that this understanding of home as a site of women's subordination did not apply to all women, and for many black women the home is (also) a site of refuge from racism (hooks, 1991). Thus feminist geography, like other forms of feminist theory, was criticised for its failure to incorporate experiences of (racialised) differences between women (Ramazanoglu, 1989; Moraga and Anzaldua, 1991) and for assuming commonalities between women. The same tensions were evident in relation to feminist research on so-called 'Third World women' (Mohanty, 1991; Radcliffe, 1994) because of the ways in which such women were often essentialised either as passive victims or as authentic loci of knowledge. Instead, such women must be seen as having multiple subjectivities and negotiating different femininities within different spaces.

By the mid-1990s there was a general recognition within feminist geography that gender not only fractured other axes of social difference, but that it too was fractured by a variety of differences. Work on the intersection of gender with other social identities (Kobayashi and Peake, 1994; Bell and Valentine, 1995; Chouinard and Grant, 1995) has meant that feminist geographers have come to engage with the category of gender not as a single or unitary category but as one that is both fractured and fracturing. While producing more complex understandings of gender identities, this diversity has inevitably provoked debates about the stability and utility of the category of 'gender' as an analytical tool (WGSG, 1997). Further challenges have been added to this picture by post-structuralist work on identity which has also challenged the notion of unitary identities (Pile and Thrift, 1995). For the purposes of our argument here, most important in these debates has been the rethinking of sex and gender.

In the work described above feminist geographers have assumed that sex is a biological category, and that gender is a social construction mapped on to that category. Contrary to this, Judith Butler has been particularly influential in arguing that gender is performative, and it is this daily repetition of gender that reproduces the fiction of binary sex (Butler, 1990). While this performance is not voluntary, it can be resisted and this allows for multiple possibilities, multiple ways of 'doing' gender, multiple ways of 'doing' sex (Cream, 1995). As we indicated in Chapter 1, these performances suggest that the embedding of the gender identities within the body is a complex social process rather than a historical or biological fact. Some feminist geographers have begun to work through the implications of such arguments by focusing, for example, on the ways in which the body might be transformed to challenge binary fictions of sex and gender (Cream, 1995; Johnston, 1998). Other work has concentrated on how bodily performances can reinforce or challenge existing sets of social relations and dominant versions of masculinity or femininity (McDowell, 1995; Longhurst, 1998). Similarly, one of the aims of this book is to examine the ways in which women both mobilise and challenge naturalising discourses of gender through contestations of femininity.

This section has highlighted some of the different literatures on identity that have shaped the ways in which we think about gendered identities. A key area of debate, as we have highlighted, is the extent to which gender remains '*the* significant difference' (Haraway, 1991), given recognition of the ways in which gender is fractured through other forms of social difference as well as recognition of the unstable relationship between sex and gender (Longhurst, 1997). Such debates have been important in ensuring that gender categories – such as 'woman' – are not reified or fixed in time and space. Instead our focus is to understand how the meanings of gender – and specifically the meanings of femininity – are constructed and contested within specific contexts. Below we return to the literatures of economic, political and cultural change identified in the first part of the chapter and reflect on how these are important in the negotiation of 'new' femininities.

Spaces of oppression or spaces of opportunity? The emergence of new femininities

Feminists in geography and the social sciences have spent considerable energy assessing how the broad set of processes we refer to as globalisation has influenced the lives of women, and how women's identities are shaped as they embrace and/or contest these processes. In turn, although to a lesser extent, feminist geographers have also become interested in how gender influences the form which globalisation takes and the ways in which its processes are negotiated in different places. In this final section we examine the gendered nature of this two-way relationship in detail. We consider the extent to which these changes represent new constraints for women's lives or may provide new opportunities for renegotiating gender identities, and we ask how women deliberately embrace and contest these processes. We do so by examining processes broadly grouped around the themes of the economic, the cultural and the political discussed earlier in the chapter. First, we highlight issues around perhaps the most widely acknowledged gendered aspect of globalisation – the 'feminisation' of the labour force. Second, we analyse how cultural change, brought about through the intersection of the global and the local, has a significant influence on the lives of women across the globe and vice versa. Third, we focus on the gender aspects of policy changes associated with the growth and spread of neo-liberalism and the changing nature of the state and political organisation. We consider each of these issues in turn, drawing out their intersections and illustrating how these processes are implicated in the construction and contestation of identity and new femininities. Through these concrete examples we emphasise the importance of bringing together the globalisation and identities literatures.

Gendered economies?

An analysis of the 'feminisation' of the labour force illustrates how the impact of global economic restructuring, and the meanings and values attached to these processes, has also been spatially and socially diverse. Globalisation gives rise to new demands for labour and new concentrations of and contexts for specialisation within and across First/Second/Third World divides. For example, in a First World country such as Britain, the rapid rise of the service sector has provided some middle-class women and men with increased access to highly paid jobs in certain sectors such as finance and insurance (McDowell and Court, 1994a, 1994b). This 'new' economic space offers women in particular social and economic advancement; nevertheless, they must continue to negotiate their gender difference in work-places that are dominated by men. McDowell (1995) illustrates how women working in merchant banking in London, for example, adopt specific forms of female behaviour to 'fit in' with existing business structures, acting as 'honorary men' in some situations but also being willing to utilise 'feminine' characteristics – such as flirting to close a deal, for example. Moreover, McDowell's study also provides a clear insight into the ways in which bodies and bodily appearance matter (though it does not challenge the assumption that biological males and females exist). At the same time as some middle-class women are gaining greater access to core parts of the labour market, however, even greater numbers are being drawn into other service occupations which are often low-paid, part-time and with poor employment conditions. These jobs are sometimes precisely those that sustain the well-paid jobs of other women and men, such as the increasing numbers of domestic workers in the UK (Gregson and Lowe, 1994). Thus, the feminisation of the labour force does not automatically lead to improved conditions for everyone even within one country. In this example, the experience of the labour market is fractured by class and gender and is not therefore universal.

In the Third World globalisation has been stereotyped by the influx of 'green', cheap female labour into a so-called 'New International Division of Labour' dominated by foreign-owned multinationals (Elson and Pearson, 1981). This characterisation has been widely challenged not only for its representation of Third World women as passive, unwilling victims of capital (Lim, 1990) but also for failing to differentiate the ways in which women in different places are drawn into the capitalist world system (Pearson, 1986). A focus on the newly developing Latin American flower industry, for example, shows that not only does women's engagement with new industries vary by region, but also that new industries embedded within and founded upon new global networks are shaped by local and specific gender identities. While a recent report called attention to the poor and dangerous working conditions in this industry (TUC, 1997), Coulson (forthcoming) argues that both the industry's working conditions and its relationship with new negotiations of gender identities vary from place to place. Conditions in the Colombian flower industry, for example, are poor, whereas in Ecuador, where flowers for export have been produced under a liberalisation programme sponsored by USAID (United States Agency for International Development), conditions for women in many cases are relatively good. Most Ecuadorian women labourers are paid above the legal minimum wage and provided with health-care, maternity and breast-feeding leave. For many of these women the flower farms represent a new space where they can work as waged labourers for the first time and negotiate their femininities and roles in their rural community from a different stand-point from the women in the Colombian farms criticised by the TUC report. According to Coulson the difference between the Colombian and Ecuadorian industries is partly explained by the ways in which new paid-work relations are embedded within and mediated by local gendered social relations. The male owners (*'patrons'*) in Ecuador are mainly drawn from old, elite, land-owning classes who now live in the cities. Unlike other Andean countries affected by successive bouts of land reform, migration is cyclical in Ecuador and people of all classes retain close ties with their rural homelands. Consequently, their business relations are mediated by the honour of their family name and notions of paternalism for their workers, many of whom have lived on or near their land for many years. Thus, in this example the feminisation of the labour force related to processes of globalisation is partly shaped by existing gender relations.

While studies of globalisation have explored the two-way relationship between processes of globalisation and the construction of gender identities, work on neo-liberalism has been far more impact-oriented. Generally, this literature has avoided analysing the ways in which neo-liberal policies are in part negotiated in relation to constructions of gender. Instead, it has largely focused on the ways in which neo-liberalism has affected different groups of people in different places. For example, the recent UNCTAD report (UNCTAD, 1997) suggests that adjustment has resulted in the further marginalisation of vulnerable groups and an increase in the gaps between rich and poor at most scales. In the so-called First World, and especially in the UK since the early 1980s, impact-oriented approaches have led to a focus on issues of social exclusion, social polarisation and whether or not an underclass has developed in response to shifts towards a neo-liberal economy.

In the Third World most literature on such changes has been written within the 'gender and development' (GAD) paradigm which focuses on showing how women have often been left out of development initiatives and countering this omission by promoting gender-sensitive policy-making at all scales. Key findings from the literature written from this perspective have highlighted that women bear most of the burden of SAPs (Elson, 1990). In Peru, for example, feminist analysis has shown how recent cuts in the public sector have influenced school drop-out rates for girls and have led to decreased standards of women's health-care (Mujer y Ajuste, 1996). Other work has show how, in times of restructuring, certain groups of women are more vulnerable than others. Elderly women, for example, are currently being affected more than men by pension reform in Latin America (Clarke and Laurie, forthcoming) and traditional sectors of women's employment are experiencing severe competition as the ranks in the informal sector are swollen by restructuring and underemployment (Tanski, 1994). An important issue for those promoting gender and development strategies since SAPs were first introduced on a wide scale in the 1980s has been the documentation of ways in which households headed or maintained by women have been disproportionately affected by adjustment policies (Owen, 1996; Varley, 1996; Chant, 1997).

Gendered cultures?

Technological change, in particular the use of new information and communications technologies (ICTs), has been one of the hallmarks of globalisation. Time and space have been compressed by rapid communication technologies and virtual interfaces. As we suggested in the first section this technological revolution both shapes and is shaped by culture. With regard to cultural identities and specifically gender identities, some authors have celebrated the possibilities that the use of new globalising technologies might offer for the reworking of gender identities. Light (1995), for example, suggests that women may well find the networking possibilities of Internet technology more suitable for their political organising than traditional formal political structures because it facilitates less hierarchical networks of communication and can be highly flexible for women to use (if they have access to it). Donna Haraway (1997) explores the scope for radically reworking gender and embodied identities which may be possible in cyberspace where women may choose a multiplicity of changing identities to use in their communication with others which may bear little relation to their 'physical' selves (in newsgroups, for example).

Yet, as we have already suggested, such technologies rarely escape the uneven social relations from which they arise (Kitchin, 1998). These uneven social relations control access to the appropriate hardware which is not only much more widely available in some countries than others, but also distributed unevenly within countries (McKinsey and Company, 1997; Motorola, 1997; Kitchin, 1998). Therefore, while some women may find opportunities to rework gender through access to new technologies, these opportunities are not open to all women, and especially not to the poor. Moreover, social relations are also important in shaping the use of ICT within different contexts. Holloway et al.'s (1998) study of young people's use of ICT in the classroom in Britain, for example, demonstrates the importance of social relations in shaping use of the Internet, as well as pointing to the potential space the Internet offers different groups of girls and boys to renegotiate their gender identities. In particular, gender differentiation is reinforced through the heterosexual economy of the classroom; however, this process does not result in the simple replication of 'traditional' femininity. Similarly, other work shows that while ICT may provide new opportunities for the negotiation of femininities, it can also be appropriated by individuals and institutions for use in negatively gendered ways, for example to sexually harass or to circumvent legislation against pornography and discrimination (Valentine, 1998). The technologies in and of themselves are therefore neither progressive or regressive; they can be used to reinforce or to challenge exisiting gender identities.

The cultural processes of globalisation might also be expected to challenge gender relations. Several studies have emphasised that in fact global migrations and the creation of diaspora populations may result in the reinforcing of restrictive roles for women – for example, the racialised gendering of Filipino women as domestic workers (Pratt, 1998) or the involvement of Eastern European and Thai women in the sex trade in European cities such as Amsterdam (Pettman, 1996). Furthermore, as we have already suggested, there are often tensions within diaspora populations about embracing cultural transformations or seeking to maintain cultural integrity and these concerns often focus on retaining assumed 'traditional' gender identities. Yasmin Ali's (1992) account of how normative views of femininity are reinforced in a poor British Muslim community is a good example.

Yet globalised cultures and mediated knowledges also open up possibilities for reworking gender identities. Parminder Bhachu (1993) traces the ways in which British Sikh women have transformed the dowry system. In its translation to the UK the dowry has become a means for professional single women to build up their own financial independence. Global cultures may also offer opportunities for reworking gendered identities, as Marie Gillespie's (1995) study of the use of the media by young British Asians suggests. Gillespie outlines how 'Bollywood' movies seen on satellite TV are reinterpreted in a local London context to renegotiate gendered expectations with regard to arranged marriages. Around these and other aspects of globalisation it is possible to begin to identify forms of gendered politics.

Gendered politics?

As the earlier part of the chapter suggested, neo-liberal policies are instituted within and by nation-states, usually with the support of international agencies such as the IMF and the World Bank among others. Because nation-states play an important role in establishing and regulating gender identities (Yuval Davies, 1997), the influence of state ideologies of gender means that the legacies of earlier gender regimes still continue to structure society, even when specific powers of the nation-state have been 'rolled back'. While hegemonic identities are usually actively contested, the legacies of different political regimes can have lasting importance. Using the case of Peruvian state ideologies of gender, Sarah Radcliffe (1993) illustrates how, during the period 1968–1990, different governments promoted contradictory contexts for the formation of gender identities. One set of government policies cast women as genderless class actors and, in particular, saw indigenous women mainly in terms of a class-based rural identity (as *campesinas* or peasants). These constructions brought women and indigenous people into a state-led project of nation-building. Another set of ideals defined women in terms of motherhood, thereby accommodating femininity to the needs of the state, whether it be a paternalistic state with welfarist policies to protect vulnerable women and children or a neo-liberal state where mothers become *'super madres'* (super mothers) (Chaney, 1979) whose collective actions in daily survival carry the nation through times of economic and political crisis. In Peru a series of state-led ideologies of mothering and shifts towards class-based constructions of both indigenous people and urban women produced a powerful, yet confusing and sometimes discriminatory, set of gender regimes, the legacies of which have been long lasting (see Figure 1.4).

The continuing influence of past gender ideologies can also be seen in the sometimes conflicting constructions of gender in the countries of EEFSU. Indeed a focus on these ideologies can usefully illustrate the links between state gender ideologies, economic relations and the implementation of neo-liberal adjustment packages. Any attempt to generalise among all women in eastern

Europe is dangerous, but broadly speaking communist states incorporated 'emancipation' as official policy. This was achieved by the incorporation of women into full-time paid employment (women as workers) while retaining traditional gender roles of female responsibility for family and home (women as wives and mothers) in what can be described as a 'state patriarchal system' (Smith, 1999b). With the end of communism, the 'transitional economies' of EEFSU have been opened up to globalised economic pressures, which at a time of increasing global competition has led to a decline in many sectors of manufacturing. What is particularly interesting in the case of EEFSU is that economic restructuring has led not to a feminisation of employment (as we saw in the case of many western and southern countries) but to a sharp fall in economic activity rates among women. This 'return to the home' is by no means universally unwelcome as it reduces the heavy 'double burden' of full-time employment and family commitments (Corrin, 1992; Einhorn, 1993), and this fact partially explains the lack of mass protest at high levels of female unemployment during the years of post-communist 'transition' (Bridger et al., 1996). The influence of state policies in failing to support female employment explicitly, and in some cases specifically promoting a policy for women of going 'back to the home' as a method of social change may make staying at home an economically rational strategy. This is particularly true in cases where costs for childcare or care for the ill and elderly outstrip possible income (Bruno, 1997). Such processes in relation to German reunification, for example, have led to some portraying women as the 'losers' of unification (Figure 2.2). However, women have also actively sought to alter the political frameworks of change, as suggested for example in the reformulation of the German constitution to include gender equality issues, and to assert the active role of women as political citizens, in the slogan 'no state without us', and as individuals working hard to deal with very difficult circumstances (Bridger et al., 1996; Renne, 1997).

What these examples show is that, even though nation-states may well be changing their roles and losing certain powers in the face of global neo-liberalism, state-led gender regimes have long-lasting and important implications for constructions of femininity. These constructions continue to be changed as neo-liberal nation-states reshape economies, redefine notions of citizenship and reorganise production, reproduction and the provision of welfare in keeping with global fashion. Whether such changes can benefit women is not clear. The higher participation of women in community and neighbourhood actions suggests some forms of reorganisation may well open up new spaces in which women can exercise control in their lives. Alternatively, access to power may move from public domains to more privatised structures where networks of power and influence, or 'old boys' networks', may dominate. Furthermore, Regulska (1998) suggests that the failure to incorporate local forms of political action in the understanding of how the whole network of power and control is being transformed during post-communist transition in Poland, for example, effectively silences the actions of many women who are actively participating in, contesting and shaping the processes of transition in local and everyday spaces. Thus, recognition of women's roles in NSMs may well open up new spaces

Figure 2.2   Women as the 'Losers of Unification'. © Marie Marcks, Heidelberg, 1992

for action where previously women had not been involved or where their actions had gone unnoticed. In the development literature, discussions of women's involvement in NSMs have focused on the importance of such spaces in enabling their 'empowerment' (Kabeer, 1994; Karl, 1995; Rowlands, 1997). While the term has come to be used frequently in policy formation by development agencies, its origins lie in grass-roots movements where according to Kabeer (1994: 229) '(t)he multidimensional nature of power suggests that empowerment strategies for women must build on "the power within" as a necessary adjunct to improving their ability to control resources, to determine agendas and to make decisions.'

In the West the most notable example of such empowerment through an NSM is the growth of second-wave feminism from the early 1970s onwards. As a movement, feminism challenged both traditional interpretations of politics

– most famously by insisting that the personal is political – and traditional gender roles, through campaigns for equal pay and employment opportunities, for the right to control personal sexuality and reproductive capacities, and against sexual harassment and domestic violence. Many successes have been won, improving the everyday lives of large numbers of women. Nevertheless, tensions within feminism as a movement have long been apparent, with minority women claiming that they were excluded from or marginalised within a movement dominated by white, middle-class feminists. Women's involvement in NSMs and other political movements, however, is not always primarily on the basis of their identities as women. Alliances around neighbourhood, for example, or around nationalism, may be deemed more important, subsuming gender interests under other concerns. Having said this, though, involvement in such movements may also be on the basis of specifically gendered identities. This is the case in nationalist struggles, where we often find women's participation defined by their role as mothers, both embodying the nation and guaranteeing its future existence (Mazumdar, 1995; Dowler, 1998). On the other hand, motherhood may also be mobilised to legitimate actions of defiance against state regimes of power through precisely the characteristics that the state itself promotes as desirable for women, as in the case of the 'mothers' of the Plaza de Mayo in Argentina campaigning for information on 'disappeared' relatives (Radcliffe and Westwood, 1996). Therefore, women's involvement in NSMs does not always challenge existing gender roles and does not mobilise aspects of femininity such as motherhood in a homogeneous way. Certain types of 'popular feminisms' around soup kitchens or childcare strategies, for example – projects around what are seen as 'traditionally' female activities – often allow an accommodation with existing gender roles and identities rather than pose a challenge to them (WGSG, 1997). To summarise, femininities based around dichotomies of seemingly 'traditional' or 'new' identities may be radicalised in a variety of ways and have different meanings attached to them. As such, they might both challenge or be complicit with existing notions of nation-building. Occasionally, as in the case of the mothers of the disappeared in Argentina, we would like to argue that they can do both.

While it is important not to underestimate the current and ongoing differentiated effects of globalisation in many countries, it is also important not to focus only on women's identities as 'vulnerable' but also to recognise their capacity to resist. This has been made particularly explicit in the literature on women's urban social movements in Latin America where the role of women's grass-roots activities in coping with economic and political shocks has been well documented and celebrated (Barrig, 1989; Jelin, 1990; Fisher, 1993; Radcliffe and Westwood, 1993; Stephen, 1997). These studies illustrate how such activities often break down barriers between public and private space and give women access to new social spaces of solidarity which can legitimate changes in gender relations and gender identities (WGSG, 1997). These urban social movements therefore engage indirectly in identity politics. Nevertheless, it is important not merely to replace conceptualisations of women as victims with one of women as 'resisters'. Seeing women only as resisters implies that women automatically embody progressive attitudes and assumes that women are seldom complicit with conservative or oppressive agendas (Laurie and Smith, 1998). Indeed, as the focus on identity politics illustrates, any conceptualisation of women that seeks to represent all women is ill-fated, failing as it does to take into account the differences between women, the multiple ways in which gender identities are performed and the complex and sometimes contradictory ways in which new femininities are forged.

Conclusion

In this chapter we have sought to outline the significance of some of the key dimensions of large-scale processes of change which are often grouped under the heading of 'globalisation', whether it be in economic, cultural or political terms, for thinking about how femininities might be changing. Furthermore, we have discussed in some detail the ways in which feminist geographers, and others, have developed an understanding of gender identities which incorporates attention to the multiple and even fractured nature of identities around gender, 'race', sexuality, class and so on. We have begun, therefore, in this chapter to explore, firstly, the ways in which changing femininities and changing geographies must be understood in relation to each other, and, secondly, how these intersecting processes must also make room for an attention to the diversity of women's experiences across space and place and across the differences of identity among women.

In the remainder of this book, then, we aim to develop this agenda of thinking about changing gender identities, and specifically changing femininities among a range of women, through an engagement with geographical processes and concepts. We examine these in detail through an engagement with four empirical case studies. In doing so, we highlight the difference that geography makes in the formation of gender identities but we also seek to understand how these geographies change, how they are differently understood, contested and reshaped. We also focus on how these geographies are themselves gendered in powerful, diverse and changing ways. The next chapter addresses how we have worked through these issues in practice, introducing the four case studies and discussing the methodologies used to analyse the geographies of femininities.

CHAPTER THREE

Working with genders and geographies

Introduction

In this chapter we turn to an examination of some of the simultaneously practical and theoretical considerations involved in formulating appropriate research strategies which focus on changing femininities. Referring specifically to our own research projects, we address how we work with the complexities of ideas and practices of 'genders' and 'geographies'. We highlight the methodologies that helped us inform our questions about the geographies of new femininities in each of the case studies which constitute the next four chapters. In formulating the research questions for the individual case studies all of us have drawn on the extensive debates in feminist geography and related fields on feminist epistemologies, methodologies and methods. Before moving on to discuss our specific case studies, we examine briefly some of the debates concerning feminist epistemologies. According to the WGSG (1997: 87):

> Feminism challenges traditional epistemologies of what are considered valid forms of knowledge. Feminist epistemology has redefined the knower, knowing and the known (Harding, 1987; Moss, 1993: 49). It questions notions of 'truth' and validates 'alternative' sources of knowledge, such as subjective experience. Feminist epistemology stresses the non-neutrality of the researcher and the power relations involved in the research process (D. Rose, 1993: 58). It also contests boundaries between 'fieldwork' and everyday life, arguing that we are always in the 'field' (Katz, 1994: 67).

This assertion of the validity of a range of forms of knowledge advances debate on the relative merits of quantitative and qualitative methods of investigation. In the main, while qualitative or intensive methods address the 'how' questions of research, often with in-depth studies, asking about meaning and debates, quantitative, extensive methods are generally used to investigate how widespread trends and processes are. Ribbens (1989), D. Rose (1993), Gilbert (1994) and McLafferty (1995) make similar distinctions between qualitative and quantitative methods within feminist research. They suggest, firstly, that there is nothing inherently feminist in either type of research, and secondly that a combination of methods often helps keep research agendas sensitive to a range of questions and issues.

Challenges to the idea of the neutral and detached researcher raise questions about the power relations and ethics of research. Debate has focused on the positionality of the researcher and researched and the possibilities and limitations of reflexive research strategies (McDowell, 1992; D. Rose, 1993; Gibson-Graham, 1994; *Professional Geographer*, 1994; G. Rose, 1997). Assertions of the validity of a variety of ways of knowing have also informed debates in feminist geography on the politics of representation and the usefulness of textual strategies that are less linear and more polyvocal (Crang, 1992; Radcliffe, 1994). Feminist interventions have undoubtedly challenged and transformed human geography, including what is sometimes called 'the project' of critical and radical geographies. However, as Chapters 1 and 2 indicate, feminist geography is in turn challenged to realise that a concern with gender provides no immunity against other exclusionary practices (Summerville, 1993; Chouinard and Grant, 1995). The challenge of working with ideas of shifting and multiple identities remains.

All of these debates have been influential in focusing attention on the ways in which epistemology, methodology and the specific methods adopted influence the outcomes of research and the types of knowledge produced. Their importance to feminist geography has been highlighted by the increasing number of books and special issues of journals dedicated to these issues (for example, *Canadian Geographer*, 1993; *Professional Geographer*, 1994, 1995; Jones *et al.*, 1997; WGSG, 1997). While these debates have underpinned the methodologies developed in the four research projects discussed in Chapters 4–7 (see Laurie, 1995; Holloway, 1996; Smith, 1996b; Dwyer, 1997), in this chapter we move the debate on by considering how some of the theoretical issues raised in Chapters 1 and 2 were (imperfectly) translated into our research practices in the case studies. The chapter uses a slightly different format from the rest of the book in as much as the arguments are presented almost in the form of a dialogue. This includes the use of long extracts from the four individual authors to explain how particular issues were addressed in the specific contexts of our different research projects. These extracts form the basis for reflection and discussion of the points we wish to raise. In this way the chapter is intended to focus on the ways in which gender and geography are conceptualised in practical ways through 'on the ground' practices.

In the first section we provide a brief introduction to the four research projects, explaining our interest in these issues and outlining our methodological approach. The next section develops some of the debates mentioned here around the themes of 'fieldwork', 'positionality' and the politics of research. We then turn to issues of working with gender and consider the different ways in which we employed notions of gender in our research projects. Finally we consider the geographies of our work and explore the different conceptualisations of space and place that informed the projects.

The research projects

In this section we provide a brief introduction to the four research projects around which this book is based. These research projects are diverse in their spatial focus and the issues they consider; however, all inform our understanding of the geographies of new femininities. Nina Laurie focuses on women's employment in emergency work programmes in Peru. Sarah Holloway examines the socio-spatial organisation of pre-school childcare provision and the importance of this in the lives of mothers with young children in the UK. Fiona Smith considers eastern German women's responses to reunification and Claire Dwyer explores the process of identity construction among young British Muslim women. Having introduced the research topics, and explained why we consider them important to the study of changing femininity, we briefly highlight the methodologies deployed. This information is necessary for understanding the debates about our approaches to gender and geographies which are considered later in the chapter.

Women's involvement in a state employment programme, Peru: Nina Laurie

My research, which was carried out part time over three years while I was employed as a secondary school teacher in a British School in Lima, Peru's capital city, focused on women's experiences in a state-backed employment scheme. Previously, I had lived in Peru while undertaking research for a Masters thesis on women's involvement with New Social Movements (NSMs), and during this time I noticed that many women volunteers involved in the soup kitchens I studied were keen to seek employment in a new state-backed Workfare programme. I became interested in this phenomenon because, while my work on soup kitchens illustrated the importance of women's NSMs in changing domestic gender relations, it also highlighted certain limitations (social movements did not provide money and were often not taken seriously by male partners). I was curious, therefore, to know what influence women's participation in Workfare would have on female solidarity and gender relations more widely. In particular, as Chapter 4 indicates, even though the Peruvian Workfare programme was meant to attract men, large numbers of women enrolled. Consequently, I questioned the influence that being paid to do 'men's work' would have on low-income women's conceptualisations of work. In theoretical terms, I wanted to scrutinise an emerging polarisation in the NSM literature: on the one hand NSMs, and particularly such 'popular feminisms', were celebrated over all other forms of activity, including paid work; on the other hand women's grass-roots activities were dismissed as part of gendered double/triple burdens and as merely short-term survival strategies. My choice of research topic, then, was influenced both by my personal experience and by my academic and policy-based questions.

I used intensive and extensive methods in my case studies of the capital Lima, and of the provincial town of Andahuaylas. I started with semi-structured interviews with programme employees (taped and transcribed in Spanish), most of whom were women, in order to identify the issues that they felt were significant about the programme.

Friends who lived in the communities acted as research assistants helping me carry out these interviews. Their advice about language and cultural issues was invaluable, and after the interviews they would often provide me with additional background information. This method was supported by other intensive research which included newspaper archive work and semi-structured interviews with programme officials as well as more formal interviews with people involved in non-governmental organisations related to gender and development and/or Workfare. Following these interviews a questionnaire survey was designed (with advice from the research assistants) to assess how widespread some of the trends identified in the interviews were and systematically to compare the two case study areas. A random sample of employees for the survey, together with information about specific projects was gathered from government archives. The socio-economic data in the archives were often incomplete and consequently basic questions covering these issues had to be included in a questionnaire. This made it time-consuming to complete and it was often difficult to maintain people's interest, especially as many women with young children answered these questions at the door of their homes.

Motherhood, mothering and the use of pre-school childcare, Britain: Sarah Holloway

The social organisation of childrearing, that is the way children's physical and developmental needs are interpreted and catered for in particular times and places, is important because it affects the position of different men, women and children in society. On the one hand the social organisation of childrearing affects the position of different social groups, and people within those social groups, through the specific form and distribution of childcare work; on the other hand it influences the type of provision different children receive. In Britain today mothers are generally held responsible for the care of their children. Depending on their situation they may provide this childcare themselves or call on a number of other people or organisations to 'do this for them'. Consequently, childcare is delivered through a complex and piecemeal system of familial, private, voluntary and state provision (Cohen, 1988; see also Meltzer, 1994). The importance of childrearing in women's everyday lives – both emotionally and as (potentially) gratifying, gruelling and time-consuming work – and the complexities of this childcare system – which distributes caring work unevenly between men and women, and among different women – fascinated me, and this interest was one of the factors that led me to undertake postgraduate research. My thesis came to be concerned firstly with the ways in which geographers do, and could, study childcare, as well as with the geography of pre-school childcare in a specific time and place. I explored these twin themes through an empirical study based in Sheffield, UK, in the early to mid-1990s.

My first approach was to examine the importance of differences in geographers' approaches to the study of childcare. I was unhappy with the liberal theoretical underpinnings of the territorial justice tradition, and set out to demonstrate empirically the importance of my concerns (see Holloway, 1998a, for further details). I therefore undertook a study of territorial justice – correlating rates of non-parental childcare provision in each ward of a city with need variables derived from the census – and compared the results with a study of childcare use, gathered through a door-to-door questionnaire survey administered in a representative sample of enumeration districts within two case study wards. The fact that this alternative geography of childcare use was still firmly fixed within the same liberal tradition of social justice, and was based on the use of quantitative methods, allowed me to produce a critique of the territorial justice tradition in the terms acceptable to that tradition. My second approach was to try to gain an understanding of why and how parents negotiate the boundary between parental and non-parental care (see Chapter 5; also Holloway, 1998b, 1998c, 1999). In theoretical terms such a focus shifts attention away from simple questions of distribution, and towards an interest in distribution and the conditions surrounding that distribution which is more common in post-structuralist work on social justice (Young, 1990a, 1990b). In order to pursue this interest I supplemented the results of the questionnaire survey with a series of semi-structured interviews with those parents primarily responsible for childcare (mainly mothers). In mixing both quantitative and qualitative methods I was able to grasp the broad contours of childcare use, as well as provide a more in-depth understanding of how and why parents used these services.

Women contesting German reunification: Fiona Smith

My research was not originally about gender at all. Studying for the final year of a degree in German and Geography as the 'revolutions' were sweeping eastern Europe and the Berlin Wall was opened, and as the momentum towards German reunification increased, I decided my PhD research would concentrate on processes of post-communist transition in eastern Germany, and in particular on how the introduction of a market economy and the politico-legal framework of the reunited Germany would affect the residents of cities and neighbourhoods that had been operating under very different conditions for 40 years. This involved close observation of changes in the urban fabric, housing conditions, town planning and property ownership. However, an early visit to Leipzig, the second city of the former GDR, gave clear evidence that many residents of Leipzig were organising locally to shape the agenda for such change (see Figure 3.1), contesting any notion that the 'transition' from communism was a matter of simply administering the move to a pre-set destination. My research therefore set out to explore this local contestation of the process of reunification, and as such assumed an interconnection, albeit a contested one, between the 'geopolitical' shifts of the end of the Cold War and local political contests around these processes (Smith, 1997). This was nevertheless explored through a series of more locally based questions: how would local residents, individually or collectively, seek to influence these major changes; how would these be received by local politicians; what strategies would be adopted; which experiences and ideologies would be drawn upon; what would be the potentially uneven and unequal outcomes of change and the contests around it?

Working through a series of extended visits over a period of two years (1991 to 1993) and using archive materials (newspaper reports, documentation, etc.) to cover the period from the mass demonstrations in Leipzig in autumn 1989 to when the fieldwork began, I aimed to gain in-depth understanding of the processes and of the

Figure 3.1 Neighbourhood action group information stand at a local festival, Leipzig, 1992. © Fiona Smith

interpretations and actions of a range of local people: residents, activists, council-lors, city officials, and so on. I combined a city focus with detailed work in five neigh-bourhoods where local action groups developed in a range of ways to challenge the types of planning and development ideas which were beginning to emerge from the city council planners and from private developers. The fieldwork used a range of tech-niques (documentary sources, interviews with activists, planners and elected coun-cillors, surveys of group memberships, observation strategies such as attending as many action group meetings, public meetings and other events as possible, and holding group discussions with members of the action groups). Most research participants were involved in a range of stages and could therefore comment on my interpretations over time. Each technique added another dimension, although I would not argue that the process necessarily got me closer to 'the' answer, but rather allowed me to explore more of the complexity of the situation.

Young British Muslim women's identities: Claire Dwyer

At the time I started to undertake my PhD there was considerable research on young Asian women (Afshar, 1989a, 1989b, 1994; Mirza, 1989; Brah, 1993) but there was less discussion of the ways in which a revitalisation of Islam as a political discourse and as a form of identity politics was important in the construction and contestation of identity for young British Muslim women. Thus though I focused in my research on the identities of young British Muslim women in general, I was particularly interested in how and whether religion was important for them in the negotiation of identity. Perhaps like all research, this focus emerged from both personal interests and broader theoretical debates. What I hoped to achieve by this focus in my research was to open up a space to study issues of identity within a framework which would incorporate both a recognition of processes of racialisation, the approach that had been developed by geographers interested in the geography of 'race' and racism (Jackson, 1987), with a feminist approach that would recognise the ways in which discourses of racialisation and gendering intersect. But I also wanted to include within these 'structuralist' approaches a recognition of the importance of religion as a dimension of both personal and 'ethnic' identity (Modood, 1988; Knott and Khokher, 1993) which had perhaps previously been ignored. I have a personal interest in religious identities and felt that they had been underexplored within feminist work (Dwyer, 1991, 1993).

In order to explore these issues I chose to use in-depth group discussions and individual interviews. My primary concerns in recruiting participants were to include a broad range of young Muslim women who might have different views about their identification with Islam and to include some young women who were not Muslim in order to explore some comparisons. To satisfy the first criterion I chose to recruit pupils from two schools in the same town. I deliberately chose to work in a suburban area, rather than perhaps the more obvious inner city sites such as Tower Hamlets or Bradford. This was partly because I felt it was important to work in a place I already knew and where I felt less of a 'tourist'. And I also wanted to see what insights I might gain from such an under-researched suburban area in contrast to work already done in Tower Hamlets (Eade, 1994) or Bradford (Afshar, 1989a, 1989b, 1994; Knott and Khokher, 1993). Using the two schools meant the young women came from contrasting socio-economic backgrounds and had some diversity of parental national background, though the majority of the pupils were of South Asian heritage which is reflective of the majority Muslim group within the UK. In the case of 'Eastwood School', which had large numbers of Muslim pupils, I presented the research outline to pupils in Year 12 and Year 11, young women aged between 16 and 19, and asked for volunteers to join discussion groups. In my presentation I did not ask explicitly for Muslim pupils and instead invited any of those present to join. This approach ensured that I gained groups of both Muslim and non-Muslim pupils largely within existing friendship groups. At 'Foundation School' a smaller proportion of Muslim pupils meant that I recruited more directly through the existing religious societies at the school, establishing one discussion group composed only of members of the Muslim society, and one group that was a mixed group comprising pupils from the Jewish society, the Christian Union and the Muslim society. The advantages of the research strategy was that I recruited individuals directly (rather than through teachers acting as 'gatekeepers') and this ensured high levels of enthusiasm for the research.

Working in 'the field' with power and positionality

All four examples in the previous section raise questions about how we might understand 'fieldwork' and, in particular, how issues of gender and other identities are involved in feminist research (Bell *et al.*, 1993; *Professional Geographer*, 1994). Within geography (including feminist geography) 'the field', as the location of fieldwork, has often been treated as a physical space distinct from the researcher's own space, a place to go to research. This is in contrast to conceptualisations which view 'the field' in political terms as something 'not naturalised in terms of "a place" or "a people"; rather located and defined in terms of specific political criteria that operate on different but connected levels' (Nast, 1994: 57). With this definition in mind, the 'field' element of 'fieldwork' for us is not just specific places (Lima, Andahuaylas, Leipzig, Sheffield or Greater London) located geographically in specific countries, nor is it only the social context of schools, neighbourhood action groups, work programmes, childcare facilities, etc. 'The field' also implies the relationships formed with people and the decision-making processes involved in developing research agendas and adopting specific research techniques. In this sense, 'a field is a social terrain' (Nast, 1994: 57) and researchers act within this terrain by engaging in (often problematic) relationships both with and as subjects of research (Sparke, 1996).

In this section, then, we discuss some of the issues of power relations and politics that shape research. Most researchers who place their work within a feminist framework would argue that some kind of political agenda informs what they do. Two common definitions of feminist methodologies have been dominant within feminist geography. The first suggests that a feminist methodology seeks to be on the side of the oppressed and invisible and in doing so seeks to 'make a difference', and in particular to make a difference for women. The second more specifically claims a feminist methodology is 'for women, by women and about women' (Gilbert, 1994). Both are aims found more widely in feminist work which often seeks to make visible the uneven conditions under which women lead their lives and seeks to make their voices heard (see Burgos-Debray, 1984; hooks, 1991). Involvement in such research can, arguably, be significant in the lives of individual women as well as affecting broader structural issues. In Laurie's research on emergency work in Peru:

many women valued the experiences they gained in the employment programme. The act of speaking about those experiences for the first time lent a political dimension to the research. It was feminist research precisely because no one else had asked them about their opinions before and many welcomed an opportunity to articulate the significance of these experiences in their lives.

However, researchers cannot necessarily know whether such aims are being achieved, nor are such aims exclusive to feminist research (Caraway, 1991; Mohanty, 1991). On top of this, research that is feminist does not escape relations of power and positionality. Like any social encounter research may be multiply constituted, never pure in intentions or outcomes. Rather than bemoaning research as always necessarily exploitative, we want to show some ways in which we found it possible, and necessary, to work with the challenges of positionality and the politics of research.

All four of us undertook our research as (relatively) privileged, white, western, educated women. Each of us dealt to some extent with researching across differences such as 'race', class, education, family structures, language and so on. These raise the key question, as Audrey Kobayashi (1994) emphasises, of 'who speaks for whom?'. Dwyer argues for her research:

in undertaking this research perhaps the most important issues for me were these questions of legitimacy and authority. My most pressing concerns as I began the research project was whether it was either appropriate, or possible, for me, as a 'white', non-Muslim woman to undertake the research. As the project progressed I came to think about these issues less in terms of the question raised by one researcher: 'Should the white researcher stay at home?' (Haw, 1996), instead recognising the extent to which such a question essentialised differences. Yet issues of legitimacy and authority remained important.

Much has been written about 'cross-cultural' research and whether or not 'white' researchers should work with, for example, black women. For some, these debates have been about methodological precision; the issue of 'matching' interviewer and interviewee, for example, or how to minimise 'race of interviewer' effects. Much more important for Dwyer were political questions about power and representation:

As a starting point it is important to acknowledge that as a 'white' researcher working with predominantly Asian women I replicated the dominant power relations structuring research on young British Muslims which has produced a legacy of 'culturalist' explanations of the lives of young Muslim women (Brah and Minhas, 1985). In a more practical way, I did not offer an alternative role model to the interviewees about who does research. Both of these issues were, and are, important to me.

Clearly, there are political questions about the representation of Others within a discipline which still struggles to engage with difference (Summerville, 1993; Chouinard and Grant, 1995). Yet, it is possible to over-simplify the ways in which difference may work in research. Mohammad (1996) and Phoenix (1994) illustrate different responses elicited through research by differently positioned researchers, yet they emphasise that there is not a 'better' or 'correct' result. Similarly, the respondents in Dwyer's study agreed that they might have responded differently to an Asian or Muslim interviewer – but they might not necessarily have given 'better truths'. Instead, what these conclusions suggest is that as researchers we must 'recognise and take account of our own position, as well as that of our research participants, and *write this into our research practice*' (McDowell, 1992: 409, original emphasis).

This position recognises the crucial questions of authority, positionality, power and representation which are bound up in the research encounter, but suggests that as feminist researchers we should seek not to surpass or overcome these questions but embrace some of the contradictions and difficulties they may raise (Gibson-Graham, 1994). This argument takes a view of knowledge not as some pre-existing and fixed 'truths' waiting to be uncovered by the researcher, but rather as *situated knowledges* produced within particular contexts and through particular interactions (Rose, 1997). Part of the process of the research design is to be aware of how the research produces specific contexts and interactions. As Claire Dwyer explains with reference to her work on young British Muslim women's identities:

I used in-depth discussion groups rather than individual interviews for the research as an important way of challenging the power relations of the research encounter. Perhaps most crucially they changed the dynamics from my asking questions to be answered by respondents, to groups of participants engaging with issues that were important to them. This did not mean that I could evade my responsibility as researcher, and I remained the conductor of the group, but it did give the particip- ants greater freedom to establish their own agenda, and indeed to choose to discuss particular issues and ignore or dismiss topics. While participants may have enjoyed the freedom of the discussion group, they were also aware of my presence and the ways in which the discussion was being conducted for my benefit. Rather than sug- gest that power relations were erased through the use of in-depth groups, I would contend that they enabled the power relations of the research to be written more explicitly into the research practice. By deciding what they would talk about, within a context where I required them to talk, both the participants and myself were made more acutely aware of the politics of the research encounter.

The transcripts yielded by the group discussions enabled an analysis that could consider the dialogic and contextual way in which ideas were expressed and identit- ies negotiated. This was another process by which I could think about writing myself into the research practice since I could consider how particular topics were negoti- ated and discussed by participants and how my own role was also negotiated. One example of this was my recognition of how one group used discussion about racism as a strategy to avoid talking about topics which they found difficult because they disagreed (in this case the extent to which some Asian women were oppressed by their husbands). The group knew that racism was a 'safe' topic on which they agreed and which positioned me in a particular way within the group. Yet understanding how this strategy worked led me to think more carefully about issues of identity and positionality in terms of my analysis. Thus the group discussion enabled me to focus not on what individuals chose *not* to talk about but *why* particular topics were difficult to talk about.

It is therefore important to open up questions about representation and posi- tionality. As Phoenix (1994) has argued, 'race' and gender positions, and the power positions these entail, do not enter into a research situation in a unit- ary or essential way. Instead she illustrates the multiple subjectivities that both researcher and researched bring to the research encounter, thus effectively mak- ing any simple 'matching' between researched and researcher in the research encounter problematic. As Mohammad (1996) argues, reflecting on her own experiences as an Asian women interviewing other Asian women, the 'insider'/

'outsider' boundary is not adequate for understanding the complex ways in which individuals are positioned and it acts to essentialise difference. The 'insider'/'outsider' debate has been raised more generally in relation to western researchers working in 'Third World' situations (Madge, 1993; Radcliffe, 1994). While some have expressed concern at a lack of reflection among western geographers and other academics working in such situations, we cannot argue that 'insider, or local academics automatically have a more sophisticated and appropriate approach to understanding social reality in "their" society' (Sidaway, 1992: 406). Rather than setting up a clear inside/outside division, Laurie argues that in fact the most important aspect for her research was the recognition that positionality, while important, was not fixed (Kobayashi, 1994; Rhodes, 1994). Her approach emphasises the role of alliances between interviewer and interviewee and shows how the positionality of the researcher can shift within various contexts and 'fields':

The contradictions of my own lifestyle, where in the mornings I lived and worked in an elitist environment (I was employed as a teacher in the British Girls' School in Lima) and where in the afternoons I experienced the opposite extreme in the shanty towns, were draining. These dilemmas, however, were important because they made me reflect on my own shifting 'positionality' as a researcher. In the context of civil strife in Peru I was frequently very dependent on the advice and protection of key informants and those people I interviewed. On several occasions these networks provided me with the opportunity to cancel interviews at the last moment because someone had warned me that an armed strike was going to be called. Economically speaking I was usually seen as a 'gringa rica' (rich western woman) but as the economic situation in Peru worsened and essential commodities became scarce I became very dependent on the informal networks established during fieldwork to obtain basic supplies such as rice, sugar, oil, milk and flour. At other times, with my connections in the school, I was a source of contacts for people, especially women looking for paid work or a market for goods they wanted to sell. Shifting power between me as the researcher and those I researched became an essential part of fieldwork.

Staeheli and Lawson (1994) summarise this approach by introducing the idea of 'between-ness':

> This recognition – that we cannot fully understand others' subjectivities and speak with authority for them – does not imply relativism and certainly must not lead us to abandon research topics. Rather, we should recognise that the space of betweenness is a site in which we can uncover the experiences and politics of marginalised groups. Situating ourselves in a space of betweeness requires us to build a new concept of objectivity that recognises the partiality and situatedness of all knowledge. (Staeheli and Lawson, 1994: 99)

Spaces of between-ness are not difficult to find – they are created every time we engage in fieldwork, not only in the particular situations of crisis in Peru, but also in situations as apparently straightforward as middle-class women researching working-class mothers in their own town. However, as Kobayashi (1994) suggests these spaces of between-ness do not remain the same: they shift, and moving with them can be the most demanding, as well as the most valuable, field experience. Ideas of research as involving a series of in-between positions imply that no single dimension of positionality covers all aspects of a research project. The positions in which research places the researcher and those involved in the research may be multiple and may vary over time and space, having complex effects, as Smith found in studying eastern Germany:

As a westerner who was not a West German, and therefore of little consequence in the inner-German power relations affecting reunification, research participants would often talk to me about 'them in the West' (of Germany) in ways which clearly did not put me in this category. Yet as someone more at home in the market economy than many of the research participants and a stranger to their history and experiences, particularly before 1989, I was often told I simply would not understand since I had not lived through particular events. However, this dual insider/outsider status was not the only position in which I was placed. For younger female respondents where I was broadly sympathetic to their ideas and political views I often found us 'sharing' experiences or opinions. With councillors of political parties with which I had little political sympathy, I found myself struggling to keep an open mind. When interviewing certain 'elites' (such as local officials) the supposed power of the researcher did not appear to be much in evidence. This was reinforced by the fact that by travelling to a different country to do fieldwork I was in a situation where I was seen as very young to be doing doctoral research. This either placed me as young and inconsequential, seeming, I think, to be someone relatively harmless who could be told information which might not otherwise have been given, or as some kind of bizarre '*Wunderkind*' (prodigy).

It is not always possible to identify any one dynamic that is most important to those involved in the research. This means it is actually quite difficult to 'write [the position of the researcher and the research participants] into the research' (McDowell, 1992: 409). However, Gillian Rose (1997) suggests that the assumption that it is either possible or desirable to achieve a level of 'transparent reflexivity', where the researcher is able to account precisely for the power relations of research, may in fact reinforce the idea that all aspects of the researcher (or indeed any other research participant) relevant to the research can actually be 'known'. The example above suggests it is possible to resist this idea and to work through moments of incomprehension as part of the research process, to admit, as Gibson-Graham (1994) says, that in certain situations she is 'stuffed if I know'. Problems of incomprehension and the imperfections of translation were more obvious for Laurie and Smith as they conducted research in another language. However, there are imperfections in how meaning gets 'translated' between the research process and the analysis in all situations. Moments of non-comprehension, of meanings slipping past each other, of even something new emerging from such uncertainties of meaning, are found whether the research is carried out between and across different languages or between different social and cultural situations (Smith, 1996a). These moments may raise significant points for analysis, as the following example from Smith's work shows:

At one group discussion in Leipzig I responded to questions about Glasgow (my home town) with some observations about what I saw as some clear parallels in terms of de-industrialisation, levels of unemployment and social problems. This effort to find commonalities, not to essentialise difference, was rebuffed in no uncertain terms by several people who said these processes were not in any way comparable. My attempts had been interpreted as belittling the severity of their experiences, or of denying the specificity of their situation.

To see such moments simply as 'mistakes', or a lack of sensitivity to issues of positionality, is to ignore what they say about the politics of naming processes as 'different' and 'similar' in particular political and social contexts. Some of us argue, then, that our knowledges are indeed 'situated' but that what has to be written into our research practices and reports is not an attempt to account for every aspect of 'our' positionality, but a recognition of the partial and often uncertain nature of 'our' and also other people's knowledges. Keeping this in mind, we now want to explore in the following two sections how we went about working with gender and geographies in our studies and how these informed each other.

Working with 'gender'

We argued in Chapters 1 and 2 that gender is not natural, but rather that it is a social construction which is constantly being made and remade. For some feminists gender is a social construction which maps onto real biological sexes, such that biological females are gendered as women, biological males as men. For others, in particular post-structuralist thinkers, gender is a social construction the everyday performance of which reinforces a binary understanding of sex, which is itself a social construction. Moreover, we argued that gender both fractures, and is fractured by, other axes of social difference. Thus, for example an individual's class identity is always gendered, just as their gender identity is always sexualised and racialised. What we have then is a concept, gender, which is always in process, being constantly made and remade, and which is at once fractured and fracturing. Unsurprisingly, perhaps, feminists have struggled with the question of how (and whether) to work with such a disrupted and disrupting analytical category. In this section we explore the ways in which we worked through this question, translating our theoretical understanding of gender into a set of research practices. Below we examine the centrality of gender in our research; we consider the difference between a focus on women and feminist research; we outline how we choose research subjects when investigating multiple axes of social difference; and how the process of research alters understandings of gender.

McDowell and Sharp argue that Henrietta Moore's definition of feminist anthropology – 'what it is to be a woman, how cultural understandings of the category "woman" vary through space and time, and how these understandings relate to the position of women in different societies' – is hard to beat as a definition of the scope of feminist geography (Moore, 1988: 12, in McDowell and Sharp, 1997: 2). Agreeing with their position, we want to argue that the exploration of the multiple meanings of womanhood in different times and places can take a variety of forms. Firstly, we want to argue that feminist geography need not always put the category gender in the foreground from the start. For example, Smith's work on restructuring in eastern Germany was influenced from the start by theoretical and methodological debates in feminist and post-colonial writings. She drew, for example, on the work of Nancy Fraser (1992) and Seyla Benhabib (1992) in critiquing the masculinist underpinnings of western definitions of 'the political'. Smith also considered the work of Homi Bhabha (1992) and others (Mani, 1992; G. Rose, 1993, 1994), highlighting the importance of the power relations and discourses of 'development' and the links between the formation of self/other identities and the multiple ways in which these might operate in colonial/post-colonial situations and other situations of domination/resistance. She did not, however, prioritise gender as a social category in her research, but rather drew on a range of possible identities for people and their actions (such as areas of residence, age, employment status, gender, political affiliation). In doing so she deliberately set out to examine how people identified themselves and their experiences of transition:

I wanted to leave it open for those participating to suggest how particular identities were more or less significant in particular contexts and at particular moments.

Thus Smith's work was informed by feminism but did not initially prioritise gender – however, as is evident in Chapter 6, her research actually tells us much about the ways in which women's lives are shaped by the processes of reunification and about the role of women in shaping these processes. This suggests that feminist approaches need not be limited to obviously gender-related topics, and that gender need not always be prioritised from the start as an analytical category in feminist work.

In explicitly addressing gender as a central concern in her work in this book, Smith is to some extent using gender identity as an *a priori* category. Specifically, here she is interested in how gender identities are changing and in the actions and politics of women in this situation. She is, nevertheless, aware of the potential difficulty of using 'women' as an analytical category:

Rather than searching for a universal answer about 'East German women', or indeed about all women, I am more interested in the range of possible ways in which women interpret their actions and changing life courses, when their experiences of combining home, family, employment and activism are in many ways very different from the experiences on which western feminist theoretical and political approaches

are based. They may, for example, reject a western-centred interpretation of their experiences. I want to avoid essentialising gender divides of 'male' and 'female' in relation to such processes by looking at the differences among eastern German women and the extent to which the category of 'women' becomes less or more significant in these processes. But I also want to remain open to the ways in which essentialist categories are used and constructed by those involved, as sources of identity or as elements of resistance.

Thus Smith is interested in the variety of women's experiences, in the ways that different possibilities of identity and different geographies of local political action are worked out, contested, held in tension, ignored or subsumed in what are seen as more important issues, such as the overall processes of restructuring between East and West, or the particular politics of neighbourhoods. In this sense the 'geographies of new femininities' in this context are a means through which to explore the experiences of the women involved and, crucially, to challenge, to re-examine and to reassess a range of theoretical and practical assumptions about post-communist transition in ways which may provide new and more fruitful understandings of how femininities 'work' geographically, and which may, in this case, provide evidence of the active contestation of alternatives to the all-too-smooth assurances of the intended outcome of 'transition'. This includes being willing to challenge 'western' theoretical positions, including feminist ones. If 'gender' is apparently very much less important than one might expect in certain situations, this may be because other identities are more important in that case, or it may be that the conceptualisation of gender and/or space which is being used is inappropriate (see Chapter 6).

It is more common, however, to focus on gender, and often on women, in feminist work. The work by Holloway on parenting and pre-school childcare provision illustrates clearly, however, that a focus on women, or issues important in the lives of many women, is not one and the same thing as feminist research. She employed two different approaches in her work, a critique of studies of territorial justice and a more in-depth analysis of local childcare cultures. Studies of territorial justice which focus on the distribution of childcare provision examine an issue that is very important in the lives of many women. However, Holloway argues that it is not necessarily feminist:

I would not argue that studies of territorial justice are anti-feminist, indeed any type of study that raises awareness of the need for childcare and access problems certainly has its merits. However, I don't think these studies are feminist because in their focus on this issue, which is often regarded as a 'women's issue', they neglect to study gender. They make, for example, normative assumptions about need – that childcare is needed when women are no longer available in the home to provide it, and to alleviate the relative disadvantage suffered by some children. They do not consider why childcare is defined as a woman's responsibility, what this means for women's sense of themselves or their position in society, or for the relations between men and women.

In contrast, her work on local childcare cultures, which is considered in more detail in Chapter 5, looks more explicitly at the social construction of gender. Specifically, she used a series of semi-structured interviews to explore the local constitution of ideas about good mothering and the ways in which these are important both in defining mothers as a social group and in influencing the meaning and experience of motherhood for individual women. Some women actively draw on these local moral geographies as a resource to inform and validate their actions; other women who might not initially choose to mother in this way find themselves under pressure from other mothers and childcare professionals to conform. Thus these moral codes influence individual women's experience of motherhood as they define themselves (and are defined) in relation to it.

The question facing a researcher who works with an understanding of gender as socially constructed, but also as a construction that both fractures, and is fractured by, other social differences is not simply whether or not to prioritise gender but also how to define one's research subjects. Dwyer chose to use the term 'British Muslim women' as the description of her research focus because it came closest to defining what she was most interested in analysing. As she outlined earlier in the chapter a key focus for her research was on how Islam – however it might be understood – was important in the lives of her respondents and what impact the revitalisation or repoliticisation of British Islam have for them. While this focus was a starting point for her analysis, she was also explicit that she wanted to consider other interrelating dimensions of identity – gender, class, ethnic background – in her research. Indeed, rather than isolating Islam as an explanatory variable, she was interested in opening up the multiple meanings of 'Muslim' which might be deployed, constructed and contested by the participants (Dwyer, 1997, 1999b) so that 'Muslim' was an open and contested signifier of identity.

Yet, clearly by choosing this descriptor Dwyer was aware of the ways in which she risked essentialising the identities of the respondents. As Lazreg (1988) explains in the context of her own research on women in the Middle East, we do not routinely define white, European women as 'Christian women'. Instead 'Muslim women' can work as a means by which 'the Other' is essentialised and exoticised, risking the danger of 'culturalist' explanations which seek to define the lives of the young women researched always within a paradigm of 'ethnic', religious and 'cultural' behaviour. As she suggests in Chapter 7 (see also Dwyer, 1999b) Dwyer tried to work beyond this paradigm in her own research – recognising the ways in which cultural and structural explanations are mutually constituted. She also tried to frame her own research within a recognition of the ways in which such essentialising discourses work to *define* the respondents, so that part of their own negotiation of identity was in opposition to these stereotypes:

In recruiting the participants for the research I tried to think through some of these difficulties of defining individuals and wanted to involve young Muslim women with a range of views as well as non-Muslim young women. Having pupils from two schools

in the same town provided a framework for this. The two schools drew pupils from contrasting socio-economic backgrounds (see Chapter 7) and included some diversity of parental national background. By using schools rather than a Muslim youth group I hoped to have a diversity of opinions about religious observance and orthodoxy. A mix of Muslim and non-Muslim young women at Eastwood School came through recruiting existing groups of friends. Recruiting through religious societies at Foundation School meant that in this case my initial approach to the participants was more explicitly through a focus on religious identity.

By this broad approach Dwyer was able to include participants in the research who had a range of different backgrounds and contrasting opinions about the meanings of Muslimness. This is a contrast to some other contemporary studies (Jacobson, 1997) which have perhaps chosen self-defined, self-consciously Muslim participants. One of the characteristics of the discussion groups was the level of debate and discussion among the participants about many different things – but particularly about what being a Muslim might mean. Dwyer acknowledges that the least successful aspect of this recruitment strategy was her attempt to incorporate both Muslim and non-Muslim girls into the research design. This was less of a problem at Eastwood School, where discussions were made up of existing friendship groups, than at Foundation School, where some discussion groups were created by Dwyer. While interesting insights were gained in these discussions, Dwyer was aware that individuals felt less 'comfortable' than in other groups and this is perhaps a particular concern when using in-depth discussion groups. In particular, individuals wanted to avoid confronting questions of 'difference' and sought to minimise conflict. While in-depth groups might usually be made up of strangers (see Burgess *et al.*, 1988a, 1988b), when working with young people different approaches need to be adopted (Dwyer, 1997).

Nevertheless, we would maintain that her strategy of incorporating both Muslim and non-Muslim participants was important in providing an effective context within which to explore the primary focus of her research since it ensured that participants were negotiating their identities within the 'everyday' spheres of their lives, even though the empirical research does not include a great deal of focus upon the non-Muslim participants. Thus Dwyer approached the construction of her research subjects through a process that recognised the dangers of overdetermining identities through one descriptor both through the mechanisms used to recruit participants, and by trying to make the category of 'Muslim' problematic even though she used it as a definition for the participants. A similar process of working with the definition of young women informed the study: the research worked with the category of 'women' to recruit participants but sought to explore how participants negotiated and expressed their identities as women, disrupting in the analysis any notion of gender as unproblematic and as somehow distinct from other aspects of identity such as ethnicity, class, age or religion. This process in turn raised questions about how the young Muslim women in the study should be represented in writing and analysis.

For example, when writing about the headscarf as a contested signifier for young women (Dwyer, 1999b) it is necessary to tread a fine line between acknowledging the *possibilities* that might exist to rework the veil as a symbol of resistance (as expressed by some of the respondents) and recognising the ways in which it remains a patriarchal practice of domination:

I have needed to ask myself in writing about these tensions, to what extent do I find it easier to emphasise, given my position as a 'white' researcher, the racialisation of young Muslim women? To what extent does my concern that I do not overemphasise the patriarchal pressures upon young Muslim women reflect the legacy of work by white, western feminism which has taken this position?

What is important in this depiction (see Chapter 7) is the tension that is created around the different possibilities of wearing the veil, as discussed by the participants. This possibility of multiple possibilities and subjectivities, within constraints, represents one approach to working with multiple gendered identities.

In contrast to Dwyer's work, Laurie's research in Peru set out explicitly to examine changing *gender relations* through the impact of involvement in Workfare programmes. What would be the effects of women taking on paid employment, particularly where the work was seen as 'men's' work? The concept of gender relations, while representing an attempt to move beyond the limitations of the idea of gender roles, still works with a particular notion of what gender is and how it functions:

> rather than construing gender in terms of socially ascribed roles, [the concept of gender relations] sees gender as a relational term, involving power relations between women and men. Although still tied to notions of male and female biological difference, it is male dominance and the processes which underlie this which constitute the main focus, [leading] to various debates over the concept patriarchy. (WGSG, 1997: 66)

However, for Laurie the actual process of doing research transformed the understandings of gender involved in the study:

Apart from the civil war, the most significant influence on my situation in 'the field' was my personal combination of full-time paid employment and part-time research in very different spaces. I was between the expatriate, English-speaking community and the Peruvian elite teachers; between life in the capital's shanty towns and relatively isolated parts of the country. This combination of activities and locations constantly reminded me that I belonged everywhere and yet I belonged nowhere. It revealed my personal contradictions and emphasised that we all hold multiple identities in tension. In order to cope with the pressures of moving in different worlds I had to form alliances with people in ways I would not have had to do had I been living in only one of the four scenarios. I had to represent those alliances constantly, in order to put people's minds at rest about the various 'others' I interacted with and also to

explain (and sometimes justify) why I did what I did. Sometimes this process of representation was very political and strategic, while at other times it seemed more like a compromise. Whatever it was, my own processes of representation fundamentally changed the way I looked at the women I was studying. Instead of looking for visible changes in gender relations I started to focus on the ways in which women in Workfare represented themselves to others. I analysed what they said about their work and themselves as employees to their husbands, families, friends and outsiders like me. I focused on the ways they drew boundaries between themselves as a group of women workers and how they chose strategically to represent certain groups of women in particular ways in order to justify what they were doing and how they behaved.

In theoretical terms, this led Laurie to focus on the ways in which the women held their multiple identities together and how they negotiated the boundaries around identities. In so doing, she moved away from a focus on gender relations that tends, 'no matter how hard feminist geographers try to nuance their accounts [. . .] with qualifying remarks about historical and spatial specificity' (WGSG, 1997: 70), to treat gender as monolithic (Laurie, 1997a, forthcoming) to an engagement with the complexities of multiple and shifting 'femininities' constituted in, and constituting a range of geographies. Clearly, then, there are questions to be asked about the connections between working with gender, the social and political relations of the process of research and the ways in which both operate in and through, create and are created by, a variety of geographies of space, place and (shifting) locations. Indeed, although we give separate consideration in this chapter to working with power and positionality, working with gender and working through geographies we hope it is clear that they cannot be disconnected.

Working through geographies

As the section on 'working with gender' has begun to suggest, our studies are less interested in understandings of space as passive, simply providing a surface on which social relations are played out, or in notions of absolute space which can be divided into discrete, knowable parts, and are more interested in relational space and the idea that place matters. We now draw on our different case studies to illustrate the ways in which particular understandings of geographies can be embedded within empirical research projects. We draw in particular on the three-fold approach to geographies and femininities first outlined in the introduction. We start by using Holloway's and Laurie's work to show how the conviction that place matters can lead to case-study research. We then focus on the ways in which Dwyer's and Smith's interest in everyday spaces shaped their examination of the importance of particular sites in the construction and contestation of 'new' femininities. Finally, we use Smith's work to consider how an interest in discourses about place, spaces and locations can be explored through an empirically grounded piece of work.

To begin, we want to use Holloway's strategic, two-fold approach to illustrate how different theorisations of space and place can be associated with different methodological tools. Holloway's first approach was designed to provide a critique of studies of territorial justice in terms acceptable to that tradition. She therefore reproduced a study of territorial justice in the distribution of childcare provision, correlating the relevant needs and provision indicators, and compared this with an alternative geography of childcare use produced through a questionnaire survey. The territorial justice approach is rooted in a liberal tradition of social justice, which assumes that goods (in this case childcare) should be distributed fairly among individuals. It works with a passive conception of space as a container of particular socio-economic conditions which should be responded to in appropriate ways such that justice is ensured. The resultant focus is on spatially just distributions. This approach has a number of drawbacks, not the least of which is that in reality space is not a passive container of socio-economic conditions, and that spatially just distributions of childcare do not, therefore, necessarily ensure social justice (see Holloway, 1998a, for more details).

Holloway's second approach was to explore local childcare cultures, in particular looking at how and why parents negotiate the boundary between parental and non-parental care. This involved building on the questionnaire survey with a series of qualitative interviews. This approach is compatible with post-structuralist approaches to social justice which, unlike the liberal tradition, do not assume that individuals exist prior to their relations with goods or individuals, but rather that they are constructed through these relations. This relational understanding of identity is combined in Holloway's study with a relational understanding of place. Place is important in the constitution as well as the experience of social relations. It is not simply a location in which people, in this case parents and children, experience a given set of social relations; these relations interweave and are transformed as they are played out in particular places. For Holloway, understanding the importance of place was one issue that shaped her decision to pursue in-depth case-study research:

In order to explore parents' decisions about childcare I had to move to the local level. This was partly because the childcare cultures I was interested in really did appear to be local cultures. Parents' decisions were undoubtedly shaped by wider social forces and policies, but these were experienced in local contexts and were often shaped through local social practices. However, the decision was also partly shaped by my desire to answer the 'why' questions as well as the 'what' questions in parents' use of childcare. To explain the patterns of childcare use I needed to supplement quantitative data with qualitative analysis, and limited time and resources meant I therefore had to work at a smaller scale if I was not to gloss over significant details.

Turning next to Laurie's work we can see another way in which an interest in the importance of place can lead to comparative case-study research. The Workfare programme which Laurie analysed was nation-wide, and despite the fact that Peru is geographically, culturally, ethnically and linguistically diverse, most of the limited literature available on the programme was based solely on the Lima experience. Laurie assumed that the opportunities created by the national programme would vary from place to place in relation to the diversity of gender identities and forms of negotiating power. In order to capture some of this diversity she selected two different case-study areas (see Figures 3.2 and 3.3) to provide a metropolitan/provincial comparison incorporating differences in culture, economy and construction of gender:

I selected Andahuaylas because gender roles in this area have traditionally involved divisions of labour where domestic responsibilities have fallen to women in private spheres. While women take a public role in agriculture, the tasks they do are often different from those of men. Generally, they reflect the sorts of gendered roles described by Radcliffe (1986) for subsistence economies in the rural areas of Cuzco where women's principal roles are planting and marketing low-order produce. I assumed that paid work would therefore be something new for these provincial women. In contrast to Andahuaylas, I chose the Lima case studies on the assumption that women's exposure to paid work and to non-traditional divisions of labour and gender relations would be greater in a major urban area. These urban areas were *'pueblos jovenes'* (new towns/shanty towns) which have a large percentage of new housing invasions, comprising bamboo shacks with limited service provision where the poorest people and most recent migrants live. The majority of the women in the Lima case study were first generation migrants from rural areas, many of them from Apurímac, the department where Andahuaylas is located. As such, these areas provided a useful comparison to the Andahuaylas samples.

Comparative case-study work is not without its difficulties. Given a limited time frame for the collection and analysis of data, a researcher undertaking two case studies runs the risk of collecting superficial information on two places rather than the detailed information necessary for in-depth analysis. Moreover, if the comparison being made is not a valid one, he or she runs the risk of gaining little insight and perhaps writing two projects rather than one. Given these potential difficulties, Laurie was careful to ensure that her provincial case study was warranted and to analyse thoroughly the similarities and differences between the two contexts. Thinking carefully about the justification for the comparative study resulted in some significant observations:

Having to justify the choice of two case studies revealed that the greatest changes in the gendering of decision-making power as a result of women's involvement in Workfare came in seemingly more 'traditional' contexts (provincial households where older married women worked). It led me to critique the ways in which much gender and development literature has conceptualised change in narrow ways (focusing on visible changes in household divisions of labour rather than on more everyday practices). It indicated how change can occur by women fulfilling seemingly 'traditional' roles as mothers in new ways, attaching new symbolic importance to old practices, thereby departing from analysis which under-valued women's domestic identities. Finally, this approach led me to analyse the ways in which gender identities are negotiated in relation to complex racist and classist attitudes (Laurie, 1997a, forthcoming).

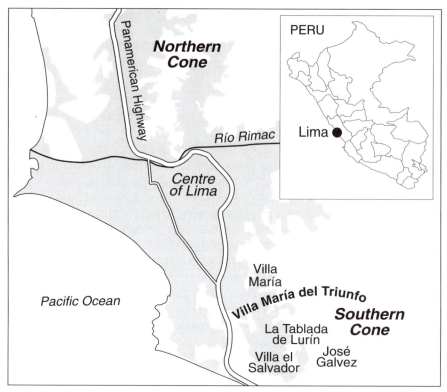

Figure 3.2 Map of Lima showing Nina Laurie's three case study sites – José Galvez, La Tablada and Villa María

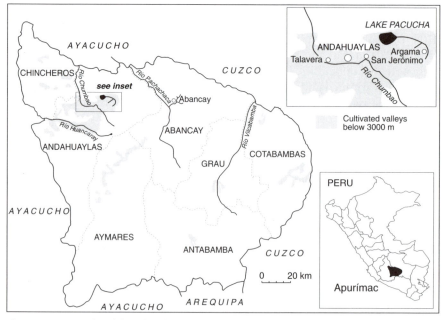

Figure 3.3 Map of Andahuaylas showing Nina Laurie's case study site

In contrast, Smith rejected the idea of undertaking a comparative study of processes of urban change in Leipzig and a western German city as she had originally envisaged. While Laurie argues convincingly that her comparison was warranted, for Smith it very rapidly became clear that the main 'comparison' would be that things were very different in many ways. Instead she came to write about the changes facing people (and in this case particularly women) in eastern Germany as events significant in their own right. This decision also has a politics to it, in that she rejected notions that see transition as being about whether 'The East' is 'catching up' with 'The West' in an 'appropriate' way, or only view experiences in eastern Germany in terms of a comparison with the 'West'.

As we have already argued in Chapter 1 the question of geographies of gender can be considered at a variety of different scales, and Dwyer is much more interested in everyday spaces:

My research considers the geography of gendered identities through a focus upon the 'everyday' spaces of people's daily lives. Thus the focus is very much upon the micro-geographies of daily life. How are boundaries between groups constructed and contested within the spaces of the classroom or playground? How are particular spaces coded as 'safe' or 'unsafe'? What kinds of gendered identities can be expressed in different places? How can alternative spaces or meanings be produced, or even imagined, through different places or at different moments?

She therefore needed to construct a methodology that would enable her to understand the dynamic and multifaceted ways in which identities are constructed and contested through different spaces in the course of people's everyday lives. It might have been possible to conduct an in-depth ethnography of the lives of the participants through participation in their daily lives. However, she chose not to follow this route for the research, as she was mindful of how difficult such an ethnography would have been to organise in any way that did not become artificial. Instead, she chose to conduct her research through in-depth discussion groups and interviews with young women at school. This means that most of the information that she had about their lives was told to her during these discussions. Additionally, she also took the opportunity while at both schools to observe the pupils during lessons and breaktimes, as well as being able to participate in some other activities such as Muslim society meetings or the fashion show rehearsals at Foundation School (see Chapter 7). However, she did not have direct access to the activities of pupils outside school. What she gained from the in-depth discussion groups, then, was a negotiated space in which to explore the meanings and importance of everyday spaces, and within which individuals could participate in the research on their own terms.

Smith shares Dwyer's interest in everyday spaces. The theoretical basis for her research was one that not only sought to explore how identities were mobilised, contested and constructed but that also rejected the notion that the spaces through which the processes of urban change would operate were ones which could be identified in advance. Instead she was interested in how these very processes would create spaces, places and scales of action, how material, political and other discursive processes would constitute these geographies in a range of ways, how the geographies which were created, contested and mobilised would have powerful outcomes for the people involved. Some examples illustrate these points:

My research was aimed at exploring the geographies of action, and the significance of different spaces for such action without preconceived assumptions that particular spaces might be meaningful in particular ways. This anti-essentialist view aimed to leave space for the experiences of those involved in the research in Leipzig. While I focused on particular neighbourhoods, I set out not to apply assumptions about the nature of 'community' politics or about the significance that local activists or others would attach to these places. The first point is important because in the German language there is no real equivalent for the type of approach to 'community' action, in terms of community as 'local neighbourhood and the people there'. I did not want to assume that people would operate in this way, when there seemed no real way of naming such processes. This has implications for women's involvement as among many writers it has been argued that women are more likely to be involved in 'community' politics, or at the local level because of the association with reproductive aspects of life rather than with formal politics and the public sphere. This is not to say that I did not examine these issues, but I was careful not to essentialise the naming of spaces and associated processes in a very different linguistic and socio-political context.

Smith's decision not to undertake a comparative study between East and West Germany, noted above, does not mean that a consideration of 'East/West' differences was not important in her study. Indeed by identifying Leipzig as being part of eastern Germany she effectively reasserts this division. However, theoretically she was interested in the contrasting ways in which such differences would be important in the ways in which transition was being worked out locally. One might argue that the view of post-communist transition as one that simply extends western practices to these areas is based on notions of absolute space which can be known, mapped and controlled, with East and West as spaces that should be purely based on a homogeneous 'western' system which is more 'developed' than that which went before in the East (Hörschelmann, 1997). Yet it is possible to think of other, more hybrid forms of spatiality in which the processes of post-communist transformation include a range of experiences, past and present, new forms of development that might avoid repeating the problems of western systems, and so on. Smith's research therefore set out both to use the notions of East and West, local, national and international, and to explore through the research how these geographies were created and contested. Was 'local' action seen as significant or insignificant for city-level actions or for broader processes of reunification? How did those involved establish meaningful spaces or contest the ways in which more powerful actors sought to limit their influence through spatial tactics and discourses? The politics of scale (N. Smith, 1993) and the powerful nature of geographical discourses and metaphors (Smith and Katz, 1993) in theoretical and in practical everyday terms informed the way in which the research was set up, and were used in turn as analytical devices to explore the dynamics of the situation (see Chapter 6).

Specifically, in relation to the gendering of these processes, two issues are of importance. The first is that of understanding how these contested geographies have implications for, are used by, are constructed by and constrain the actions of women in seeking to establish their identities in a period of change. The second is to understand how the discourses and practices of change are themselves gendered, often in ways that operate on general ideologies of gender identity but need not necessarily (though often do) affect women and men in different ways. To be more specific, the ideological association of the local with feminine characteristics, for example, may well affect the relative significance attached to both male and female participation in these actions by more powerful actors.

In summary, although we have adopted a range of geographical strategies (including comparative work, in-depth case studies, addressing major structural changes and the microgeographies of everyday life), all four studies examine both geographers' conceptions of these spaces and, fundamentally, how researchers and research participants work with a variety of often competing notions of 'space', 'place', 'location' and so on. We have also stressed that while the representations of spaces, places and locations are crucial, such representations arise from, reflect, contest and shape the material realities and social relations of the issues such as employment, the family, childcare, housing conditions, rapid political change or civil war which are involved in our research projects.

Conclusions

In this chapter we have raised key questions about the politics of research and working with gender and geography. After introducing our four case studies, we discussed a range of ways in which the important ideas of positionality and reflexivity influenced our work. We also suggested that the imperfections of such positioning and reflexivity in practice and the in-between spaces of research should be seen not as failures to implement 'pure' feminist research but as research moments which, in their very disruptions, offer productive ways to understand the research process. We then moved on to explore the variety of ways in which our studies sought to work with 'gender' and in particular how we sought to work with the multiplicity of identities that were of importance to the different situations discussed. Finally, as feminist geographers, we argued that these issues intersected in a variety of important ways with the spatialities of our research – where we researched, how space and place were defined, what notions of space were employed in the work – and that these very spatialities influenced the findings of the work. Throughout these themes we have sought to illustrate a range of key themes for this book: how space and identities intersect; how we as researchers and those involved in the research as participants understand what research can and should be aiming to do; how our and others' intentions are imperfectly implemented in practice; and how the changing experiences of practice in turn shape and reshape the substantive and theoretical outcomes of research. These issues should be borne in mind when reading the following four chapters which explore each research project in turn. In them, we illustrate some of the issues raised here about the links between geographical and feminist theories and practices through a consideration of diverse geographies of new femininities.

CHAPTER FOUR

The shifting geographies of femininity and emergency work in Peru

NINA LAURIE

Introduction

This chapter is the first of four case studies exploring the specific ways in which femininities emerge differently in different places. In this chapter the focus is on the construction of femininities in the context of state-led employment based on emergency work/Workfare initiatives (that is the provision of some form of 'welfare' in return for labour from marginal sectors of society). Globally such large-scale employment initiatives have become increasingly important as social cushions to help smooth over processes of economic restructuring and crisis associated with some of the aspects of globalisation outlined in Chapter 2. In particular, this chapter focuses on Peru and analyses the extent to which Peruvian emergency employment represented a new space for women and a departure from the existing ways in which femininity has been politicised through motherhood during the past three decades in the country. It illustrates how low-income women's identities are fractured and argues that femininities are constituted and made coherent through certain understandings of geography. It draws on work more widely focused on the feminisation of emergency work in Latin America and the role of the state and social movements in the negotiation of gender (Laurie, 1995). The research was based on survey, interview and archive work in three communities in Lima (La Tablada, José Galvez and Villa María) and in the provincial town of Andahuaylas (see Figures 3.2 and 3.3). The chapter shows how, by taking up emergency employment, low-income women in Peru gained access to new spaces while reclaiming and re-evaluating mothering identities and using, blurring and superseding binary divisions of space.

Crisis and emergency employment in Peru

In the mid-1980s, and into the early 1990s, Peru experienced one of the worst economic crises in recent history. The effects of long-term un(der)employment, slow (and in some cases, negative) growth in gross domestic product (GDP), and huge price rises in everyday commodities became the norm in the country

Figure 4.1   Defaced propaganda graffiti for PAIT, La Tablada. It shows PAIT's name crossed out and replaced with COOPOP, which was what the programme was called for the first two years of Fujimori's regime (1990–1992). © Nina Laurie

for almost a decade prior to inflation peaking to levels of 20,000 per cent between August 1989 and August 1990. While the Peruvian crisis was the worst in the continent, it reflected a general stagnation as the 1980s became known as the 'Lost Decade' for development in Latin America. During this crisis, Alan García, the candidate of the American Popular Revolutionary Alliance (APRA) party, one of the oldest parties in Peru, was voted into power. He had a radical, new economic policy which, among other things, included pump-starting the economy via state-led job creation schemes. Consequently, in 1985 the Peruvian state embarked upon a huge, nation-wide emergency employment programme aimed at stemming the economic crisis by giving emergency jobs (and therefore buying power) to the poorest of the poor. This programme, known as PAIT (Programme of Support and Temporary Income), lasted the full five years of García's presidency and continued, under another name COOPOP, into the first two years of the government of Alberto Fujimori whose newly formed Cambio Noventa party came into power in Peru in 1990 (see Figure 4.1). For the first two years of García's government this programme was economically successful as gross national product (GNP) and production figures increased and poverty levels improved. However, a second aspect of García's programme was to limit international debt repayments to 10 per cent of GNP. In opposition to this policy the international community (the IMF, World Bank and international lenders) withdrew their support for Peru, thereby isolating Peru from international finance and precipitating its economic failure. Consequently, the trajectory of PAIT was adversely influenced by global financial policies after 1987. (See further reading sections at the end of the book for more information on these themes.)

García's emergency work programme was similar to the job creation schemes that first emerged under Roosevelt's 'New Deal' in the USA in the 1930s. More recently, these sorts of massive state employment programme have become associated with welfare restructuring in countries that have made social security conditional upon participation in 'Workfare' programmes (see Further Reading). PAIT was designed to attract the urban male unemployed and pay the equivalent of the legal minimum wage. It provided employment in the form of a formal three-month contract between the state and the employee. All employees appeared on an official pay role and wages were paid by a monthly cheque. In this sense, even though PAIT was short-term work (and therefore paid no social security benefits) it constituted 'formal' rather than 'informal' employment because it required fixed working hours and wages and was bound by a contract (Laurie, 1997a).

The work carried out in PAIT was designed for men. It involved constructing access roads, irrigation channels, walls and foundations, reforestation, installing basic sanitation and painting frontages in schools and health centres. It was productive work carried out in public spaces such as streets, *plazas* and public buildings. Employees were organised into gangs comprising approximately 40 workers and specific tasks included brick and pavement laying, plumbing, making floors, surfacing sports grounds, plastering, painting, planting trees, digging holes and ditches, moving rubble and carrying out heavy-duty cleaning. Because it was hard, physically demanding, manual, low-technology work, labour was used intensively and employees frequently formed human chains to excavate heavy material and transport large loads. However, despite the fact that PAIT was designed for unemployed men, what makes the programme important in the context of this book is that much to the APRA government's surprise nearly 80 per cent of those who enrolled in PAIT over the 1985–1990 period were women. This chapter will explain this feminisation and explore how women's involvement in PAIT affected the construction of femininities among low-income groups in Peru.

The feminisation of PAIT

There are a number of explanations why work that was originally designed for men was taken up by women. Firstly, upon taking power the new government assumed that large-scale, male unemployment was the main symptom of long-term economic crisis. At this point (1985), however, unemployment *per se* was not the main structural problem but, rather, underemployment coupled with the socio-cultural expectation that men would earn a 'family wage'. Consequently, PAIT, which gave no job security and paid the legal minimum wage (with no social benefits) in a monthly cheque, was seen as 'bad work' for men who were expected to provide for their family's daily needs with their wage. Secondly, even in the context of underemployment and low wages, many men with families preferred to stay in poorly paid permanent jobs in the formal sector (because they provided social security benefits) rather than give up permanent employment to work on short-term contracts in PAIT (which paid similar wages but had no benefits). Thirdly, many men were loath to enrol in PAIT because it required them to work for a month before receiving any income and hence more immediate security could be found in the informal sector which paid a daily cash wage rather than a monthly cheque. Consequently, in the context of underemployment, rising prices and low wages, men doubled up in informal sector work or complemented a formal job with informal activities, such as taxiing or buying and selling, rather than deciding to leave their existing employment to work 9 to 5 in PAIT.

The feminisation of PAIT, however, is explained not only by male reluctance to enrol in the programme but also by women's willingness to do so. One of the reasons given for the high level of female enrolment was the fact that the employment was local. Employees were recruited from the areas where the work was to be carried out. Consequently, women who enrolled did not have to travel long distances and could return home relatively easily. There are, however, also other socio-cultural factors that explain women's enrolment.

Traditionally in Latin America, concepts of a male 'family wage' have gone hand in hand with constructions of femininity that draw on ideologies of motherhood and *marianismo* (notions of self-sacrificing women based to some extent on particular understandings of Catholicism; see Chapters 1 and 2). As a consequence, despite recent increases in overall female labour-force participation (Pearson, 1986; Scott, 1994), female labour has largely been seen as flexible (able to move in and out of the job market) and secondary to men's employment. For example, Moser (1981, 1993) indicates that low-income women in Latin America join the paid labour-force in times of crisis but leave paid employment once the crisis passes. In the context of the offer of PAIT employment and the existence of male underemployment, many women entered the programme as flexible labour in search of a secondary wage and a short-term economic solution to their immediate household needs. This explanation of female enrolment is borne out not only by the large numbers of married or cohabiting women who entered the paid labour force for the first time via PAIT, but also by the work patterns of large numbers of female heads of household. These women joined PAIT but were obliged to keep up secondary jobs (usually in the informal sector because there are few formal jobs for low-income women in Peru) in order to supplement their income and to ensure that they, as women, could provide a 'family/household' income.

What is interesting about women's initial desire for a temporary, secondary wage is that many women ended up working in PAIT semi-permanently for five years on consecutive contracts. Equally interesting is the fact that, after 1987 when the economic crisis in Peru worsened, many households in effect became 'women-maintained'. These households became almost solely dependent upon women's PAIT wages to support their families while male members of the household spent time searching for work rather than actually being engaged in employment owing to the collapse of traditional sources of male employment in the building industry and in factory work (for more details of this argument, see Laurie, 1997b).

By December 1987 the total number of PAIT employees nationally had risen to 531,456 (Bernedo, 1989) and of these 80 per cent were women. This massive incorporation of female labour into state-backed employment jarred with the domestic construction of femininity which had already become highly politicised in the 1970s/1980s in Peru. The politicisation of women's domestic role mainly occurred in the Lima-based soup kitchen and 'glass of milk a day' movements (Barrig, 1989). These well-known movements grew up as survival strategies for low-income women in the face of severe economic crisis. They represented a grass-roots response to increased food prices, declining nutrition levels and high rates of poverty in the shanty towns of Lima. The 'glass of milk a day' scheme organised committees of women to supply food-aid milk to young children while the soup kitchens were a variety of communal dinning rooms, the majority of which were located in local parishes and dependent upon food aid channelled through CARITAS (the Catholic Relief Agency) and other NGOs (Vargas, 1991). '*Olla comunes*', spontaneous common cooking pots, were also important as neighbours often brought food and cooked together for a short time during periods of severe price rises.

By the time García came to power in the mid-1980s the milk and soup kitchen movements had become extensive. In academic circles they were being identified as New Social Movements (NSMs) (Escobar and Alvarez, 1992; Escobar, 1994) because of the alliances they forged with national and international organisations, particularly middle-class feminist groups which provided consciousness raising and leadership training for grass-roots leaders. These alliances were helped by the wave of interest and funding generated internationally by the UN Decade for Women (1975–1985). Consequently, by the mid-1980s the milk and soup kitchen movements had expanded into a massive women's movement in Peru which attempted to cut across distinctions of class and ethnicity in order to incorporate issues such as human rights, health-care and income generation as well as food and milk provision.

According to Patricia Ruiz Bravo, a feminist academic in Peru, the women's movement became the strongest critical voice in Peru in the mid to late 1980s, and was the only social movement powerful and organized enough actively to denounce the economic and political crises and the deepening civil war being waged between the Shining Path and the state throughout the country in the 1980s (Ruiz Bravo, 1998). The basis of this social movement was women organising as 'mothers' and the over-arching construction of femininity was a domestic one.

PAIT: new spaces for women?

In 'going out to work' in PAIT women 'left home' and domestic constructions of femininity to engage with a series of new spaces: new social spaces and new relational spaces. The new social spaces reflected the new social environment that women came into contact with in PAIT when they worked alongside many other women from a variety of different backgrounds, all of whom had a similar economic need. Relational space refers to the fact that women entering PAIT occupied new spaces in relation to men. By enrolling and working in PAIT women used public space in new ways. They joined the previously male-dominated economically active population and did work that had originally been intended for men. While the physical spaces (such as public roads and buildings) were not new, women's engagement with them was and therefore the social construction of those spaces was gendered differently. They became spaces where women, rather than men, worked.

PAIT provided new social spaces for women because the programme drew in women from a wide range of different places, local communities and social backgrounds:

> There were young and old women there, many of us who worked were single mothers. There were also some who were widows, some who weren't and some who were married and really only worked to help their husbands out.
>
> (Woman labourer, La Tablada)

There were also women with different educational backgrounds:

> There were women who had completed secondary school, some who had given up half way through in the third or fourth year, others who had only finished primary education and some who didn't know how to read or write but they all worked together, what I mean is that it was a very diverse gang.
>
> (Woman labourer, José Galvez)

For many of these women working in PAIT was the first time they came into regular, close, contact with women who were not like themselves:

> After a while you realise that you meet every type of person there. Mainly you meet women with different ideas from yours who come from different places. [In PAIT] there are submissive women and also women who actually like swearing. So as they work, all of these [differences] are intermingled. Those women who were submissive see all of these things and possibly think that those women are doing the right thing, they begin to think that it's good fun or that it's normal, others retreat because up until then they have had a life that's very different from what they are seeing.   (Woman labourer, José Galvez)

As this quotation suggests, this new social space provoked a series of questions about women's roles and identities which are not only constituted through these new spaces but which in turn influence what these new spaces look like. It is important to recognise, however, that different women experienced these changes in different and sometimes contradictory ways. For some women for whom 'home' was not a place of protection owing to domestic violence, joining PAIT and getting 'out of the house' was a way of avoiding danger. For example, after describing the persistent beatings she received from her husband, one woman, Señora Suárez, claimed that the women in PAIT acted as her only escape:

> The atmosphere with the señoras was really friendly, you had a distraction it wasn't like being shut in your house, like that thinking about things. No it was more fun, it was an escape.   (Sra Suárez, labourer, José Galvez)

For other women, however, leaving home was dangerous as they went to work in new, hostile environments, in isolated places which involved difficult journeys to work:

> There's a path that goes out from José Galvez, we used to meet there. We all had to be there at 7 am on the dot. At 7 am we'd go. Six of us used to walk. We went to keep each other company because that place is dangerous. . . . One

time they [male plural] got me, they surrounded me, me and a friend – completely surrounded us. They robbed us with a knife. They were big men and they hit the children a lot. I fell onto the rocks while I was running away. While trying to escape, I fell over. So then we had to come and go in groups because it was dangerous for us. (Woman labourer, José Galvez)

Domestic constructions of femininity meant that women were not supposed to be in such isolated places and were 'out of place' travelling to paid work, especially employment designed for men. Therefore women were vulnerable and open to attack in these spaces not only because the places were isolated but because of the ways in which space and 'travel to work' were gendered. By using isolated places to travel to (men's) paid work women were performing 'inappropriate' femininities in highly visible ways. This visibility was also reinforced by the fact that the work was located in public places like roads, hospitals, schools and *plazas*, spaces where women were not used to being:

Señorita, I started work and all of us [*nosotras*] were working in a public road, here in Avenida Pachacútec, imagine! (Woman labourer, Lima)

The image of the gangs of up to 40 female labourers engaged in hard physical labour made women's visibility in these new spaces more evident, particularly as these were spaces they did not normally occupy:

We went up the hills and excavated the rocks from up there. Then the other group made a sort of chain, passed the rocks from one to the other and then placed them on the road: rocks on top of rocks. Then on the rocks we put small stones so that the lorries carrying water could get up to the shacks.
(Woman labourer, Lima)

The ways in which PAIT located women in new spaces had a variety of meanings for different women. While for many women PAIT was their first experience of employment, others had already experienced some form of paid work. For some women PAIT was the first time they had engaged in paid work since they married, yet for others, many of whom were female heads of households, PAIT merely represented their latest job after a string of informal sector experiences. For all these women, however, whether they had paid work experience or not, PAIT provided them with a qualitatively different experience as workers.

Firstly, the PAIT experience was novel for women because they did work designed by the state for men. PAIT therefore had legitimacy as an occupation, while being work that most of the women in the programme continued to identify as men's work:

Well for me, since I started working, we have made tracks and roads. We worked like the men in order to improve things. (Woman labourer, Lima)

It was OK I worked very well but afterwards they changed the PAIT bosses and they put in other methods of working so that we had to work like men who work gripping pickaxes and those sorts of things. (Woman labourer, Lima)

Consequently, in PAIT women realised that they could 'enter male space' and work as hard as, if not harder than, a man:

> For us there was no such thing as hard work, *señorita*, because in the *campo* [open-ground/countryside] we practically climbed up with crowbars and dug out the rocks. We carried some enormous rocks. So, *señorita*, we worked harder than a man.   (Woman labourer, Lima)

This was a shock for women as well as men. Men participating in the PAIT programme also felt that many of the tasks were not 'women's work', as one male employee from Lima explained:

> Look, for me it hit me a bit hard because you could say that the women worked too much. The tasks were very heavy, for example they carried sand, they carried bricks: they did the work of a man.   (Man labourer, Lima)

Secondly, women engaged in 'formal' employment – they appeared on payrolls and were paid in cheques, which required them to go to the bank to claim their wages. They worked fixed hours and had to practise specific divisions of labour. Consequently, one woman likened the experience to working in a factory:

> In PAIT I learned all about responsibility. For example, to work in a factory you have to work to a fixed schedule, with a starting and finishing hour. There [in PAIT] we had to do the same. We had an hour when we had to start, a time for rests and an hour to leave. More than anything it was about co-ordinating things with other people [the division of labour] to agree upon 'you doing that and me doing this', that's the same as it is in a factory. In that way 'if you do that, I'll do this', the work is the same in a factory 'you put that on, I'll put this on'. The only differences were that we worked with the land. The work in the *campo* was about handling rocks whereas in a factory you work with processed materials.   (Woman labourer, Andahuaylas)

Women entering the economically active population in this way for the first time represented what Pearson (1986) has called a 'green labour force'. PAIT women were 'green' because most of them had no previous experience of paid work and even those who did were new to the formal contract working relations outlined above. Consequently, doing 'men's' paid work in PAIT constituted a new form of socialisation in the lives of female PAIT workers.

Multiple and fractured identities

When women rather than men enrolled in PAIT, not only did each have an individual work history (as illustrated above) but also individual gender identities were constructed differently. While the gender identities of low-income Peruvian women in PAIT were influenced by state ideologies of gender (see Chapter 2) and the politicisation of certain domestic roles, a series of specific and individual experiences cut through and across these identities. Ethnic origins, life course position, personal migration histories, political party affiliation, the location of the communities they were living and working in, and whether or not they lived with male partners, parents or in female-headed households – these were all important to the way in which different women experienced PAIT. Consequently, personal expectations and moral and/or religious codes distinguished different women in the programme and once women joined PAIT many of their codes were called into question. Collective and individual tensions emerged in particular around two sets of polarised gender identities: the binary construction of 'good' and 'bad' women, and clashes between being 'mother' and 'worker'. PAIT women had to negotiate being both 'good mothers' and 'good workers'. This negotiation was difficult, not only because of the double and triple social burdens entailed but also because the worker role contradicted notions of how to be a 'good' mother which for 'respectable' low-income, urban women centred on being faithful, self-effacing, full-time wives and mothers.

In general the dichotomy of constructions of 'good' and 'bad' women drew on issues around women's sexuality. Fears about women becoming 'bad' because of their participation in PAIT largely emanated from male partners, as the following examples illustrate:

> He didn't want me to go because he had heard a lot of things. Where he works in the south there's a lorry that picks them [PAIT women] up and he said that they hassled [*fastidiaban*] the men. So, when I told him that I was going to work in the *campo*, he said that they are the worst sort of people because he's seen them go by and hassle [*fastidiando*] the men almost as if they had been selected because of that very quality.   (Woman labourer, José Galvez)

> He thought I'd find another man there . . . well there were all sorts there – housewives, and then some who were a bit lively and who said things they shouldn't say – and my husband said to me 'if you go you'll learn those things, it's better that you stay in your house [*en tu casa*]'.   (Woman labourer, La Tablada)

Rumours about 'bad' sexual behaviour among the PAIT work-force were rife and women who considered themselves to be respectable, married women talked about such activities in shocked tones:

> Some of *them* just came to flirt with the bosses.   (Woman labourer, La Tablada)

> Sometimes they just went to hassle the men and they didn't do any work, they just encouraged the men not to respect them.   (Woman labourer, José Galvez)

> They just used to flirt and laugh and they encouraged the men to take them over the back to be with them behind the shacks instead of working. When the men didn't take any notice of them they would call them 'queers' [*maricones*]. In my group there were five girls, they mainly went off with the young men, and now I see them with their pregnant bellies.   (Woman labourer, José Galvez)

The link between sexual relationships and the work-place was condemned even more strongly when the scenario involved married women. Hence, an adulterous woman was given most blame for the bad reputation of the PAIT work-force. She was censured because she allowed *(dio plaza)* the advances of the men with her generally loose, vulgar behaviour. In the words of one woman from La Tablada, 'at the end of the day the man is a man and a woman is left dirty'. Many women described the unfolding of this adulterous scenario and its implications for PAIT's image:

> *Where does the bad image of PAIT workers come from?*
> There were some girls who joined PAIT who were very lively and badly behaved. They came into PAIT and they got to know a young man and fell in love with him, usually he was married to someone else. They'd get to know each other better and better until she became pregnant and that's when the talk began: 'you've been with my husband' or 'yes my wife led me a song and dance'. All of that together with the gross language they used, it was no wonder that people thought bad things about them.   (Woman labourer, La Tablada)

In most cases when sexuality was raised as a topic it was in the context of heterosexual flirtation and extra-marital affairs, as discussed above. However, despite the general silence around other expressions of sexuality a few women raised the issue of homosexuality by claiming that PAIT attracted, and gave a space to, women of a homosexual orientation. It is not possible to make any real assertions about how widespread lesbianism was in PAIT because cultural taboos meant that it was not widely discussed in interviews. The issue was raised twice, however, once by a psychologist employed to work with the female labourers in PAIT and another time in conversation with three female employees in José Galvez who were key informants throughout fieldwork and who claimed they tried to avoid the lesbians because they were different from '*nosotras las señoras*' (we married women). In a formal interview the psychologist gave her opinion about lesbianism and the reactions of the work force:

> I knew the situation with the lesbians well. They acted like men and tried to win the attention of the pretty girls, the ones with the nice figures. To begin with the girls didn't give in to the attentions and invitations of the lesbians but little by little they did, it was as if they had a magic power to get what they wanted. I heard the comments of the mothers who were very worried as they worked there too, often together with their whole family.
>
> I suppose this threat always existed in the *barrio* [neighbourhood] but they took advantage of the situation because there were usually one or two lesbians in each *cuadrilla* [work team of 40] having close contact in small groups. It didn't take them long to win over a girl, not even five years or so but just a matter of months since getting to know them. Even now I see at least one of the couples together.   (Female psychologist working for PAIT)

While the material in this quote uncritically attributes special, almost magical, powers of seduction to 'the lesbians', and reinforces negative dualistic stereotypes about male and female sexuality, importantly it also suggests that the feminised PAIT work-place offered a space for the performance of more than just heterosexual sexualities.

What is interesting about all these various constructions of 'good' and 'bad' women is that the dichotomy was also conflated with notions of 'good' and 'bad' workers. In this scenario, a good respectable woman was hard-working, while a sexually immoral woman was considered a 'bad', lazy worker, someone who did not take her work very seriously:

> Some people said that the women fell in love, that they lost their virginity. Some said that it was just like a whore house [*alcahuetería*]. It was like a whore house because instead of coming to work they just came to flirt with the men but slyly, not in an up-front way. It caused problems within the group because the boss said that they shouldn't work like that because the gossip was causing damage. The ones who went along to be lazy were the ones who got into that trouble.
>
> (Woman labourer, La Tablada)

> I knew of a case of a girl who was married and she had a baby. She fell in love with someone else in PAIT and she had a baby by him and so her husband left her. When that happened she then separated from the father of the second child because he also left her and so finally she was left with two children from different relationships and now she's with someone else with whom she has a baby. I think this is because we are sick in the head. It's the result of joking with male friends that these things happen to us and so all of this happens because individual women aren't more serious in the work-place.
>
> (Woman labourer, La Tablada)

Strategic uses of binary categories

The binary of 'good' and 'bad' PAIT women was deliberately used by some women to allay male partner's fears about the corrupting influence that PAIT could have on 'good' wives and mothers. In the language they used, women reproduced binary 'good' and 'bad' stereotypes while firmly locating themselves within the 'good' group:

> Some people got annoyed, well I did when some people got on with it and others didn't. For example there were some *señoras* who would arrive and sit down or for a short time they would carry two spades and nothing else. The *capatazes* [overseers] would watch them and ask them to do more, then when it was our turn we would do it as it should be done and put the sand to one side, then they would get annoyed. Then we'd begin to fight about it but we would try to avoid it by each one sticking to their side.   (Woman labourer, Lima)

In this way 'good' women (workers) used physical distance to keep themselves separate from 'bad' women. 'Good' women also said that they did not join up with bad, vulgar women; instead they kept themselves separate:

> They had a different way of life from the type that one was accustomed to, some had a very vulgar mouth and had lived loose, liberal lives. They didn't think about things. We didn't join up with them, we didn't join up with the very lively ones and we tried to separate ourselves and from other groups. They used gross

words like *'no jodas mierda'* [don't fuck me you shit] which are horrible words
and they had a very brusque way of speaking, using slang like *'cojudo'* [moron].
They used these words among themselves and when they passed by each other
they would say things in a very vulgar manner *'estas hueveando'* [you're wank-
ing around] or when they weren't working they'd say *'Com'on moron estas
hueveando'*. That's how they would talk.   (Woman labourer, Lima)

'Good' women emphasised that they were not used to behaving like this or
even witnessing these sorts of things and consequently the only way they could
cope with experience was to distance themselves from the vulgar women. As
one woman said:

For those of us who were not brought up like that it was a shock so we dis-
tanced ourselves from that group.   (Woman labourer, Lima)

The women in PAIT therefore used distance and physical, spatial separation
to communicate the way in which they structured the 'morally dangerous' PAIT
work-place in order to keep 'their' femininity intact.

While the binary division between good and bad was used strategically by
women the conflation between the mother–worker tension and the 'good'/
'bad' dichotomy was also used to defend themselves against gossip based on
sexual immorality. Women defended themselves against being cast as 'bad
women' (and bad workers) by appealing to notions of being a 'good' mother:

He heard a lot of comments but it depends on individuals. Sometimes there are
women who are real jokers and we give our trust away, because there are also
men who work there and some times some women end up doing what they
shouldn't. That's why I said to my husband that it depends on each person, and
if you can make them respect you, and that wherever I can I will go to work
but if you don't make them respect you . . . and if you don't believe me go and
see for yourself. That's what I said [laughs] but at least I went with my little baby.
                                                      (Woman labourer, La Tablada)

They criticised us, yes, because men always say 'you are going to do . . . all sorts
of things in PAIT just like they say you will, you will gossip and speak about
things that are strange to your lives'. But I know that they wouldn't say anything
about me because my daughter went to work with me.
                                                      (Woman labourer, La Tablada)

Women also used their role as mothers to justify their participation in paid
work, particularly as many husbands felt that women were abandoning their
domestic femininity and responsibilities by working in PAIT:

There were tensions between me and my husband because he didn't want me
to go. He told me not to neglect [*descuidar*] my children because he was also
working.   (Woman labourer, La Tablada)

So women appealed to notions of good motherhood to construct themselves
as good women (and good workers):

> The little [money] he [my husband] gave me was for me but what I wanted was to buy something for the kids. Just at that moment that year we were coming up to the rainy season and I needed to buy boots for my children. That time they cost 2,500. I had enough for everyone because they paid us 10,000 and I was pleased because I also had enough for a kilo of sugar.
>
> (Woman labourer, Andahuaylas)

Women's arguments focused on emphasising that their motivation for joining the programme was to buy things for their children. Consequently they fused the two categories of worker and mother in order to justify their participation in paid work. Such arguments allowed women to negotiate with their male partners and to show that it was not a desire to neglect their children but the need to provide for them that made them enrol in PAIT. As the same woman explained:

> The same necessity makes one go [to PAIT] because you don't have anything to draw on when the prices rise, the number of children increase, those children want more and as they grow up they want to dress better, etc. That's why I wanted to go, I wanted to buy them boots and I didn't have any money. The first month I bought pure wool for their jumpers. I bought wool and at that time I think it was 1,500 each ball. My husband said 'It's up to you, but don't go'. 'No, I want to go'. That's what I said, I told him I wanted to work, even now I say I want to work.   (Woman labourer, Andahuaylas)

Some husbands were won over by such arguments. One husband explained his wife's participation in the following way:

> At that time all my children were studying, therefore we needed money. She saw that my money wasn't enough and so she said 'I'm going to give you a hand, I'm going to help you but I'm going to work in PAIT'. I said 'Are they paying that much? Where are you going to work?' She replied 'Here in the Zona Nueva'. Various of her friends and acquaintances had gone so I said 'Yes, all right, if you all go accompanied'.
>
> (Sr Jiménez, husband of female labourer, Lima)

Therefore, the conflation between the mother–worker tension and constructions of 'good' and 'bad' women served to allow 'good' women to locate themselves respectably in the programme as workers.

Two other factors were also important in avoiding tensions appearing in women's multiple identities. First, the most obvious reason for the lack of conflict between mother–worker identities was the fact that the severity and long-term nature of the economic crisis meant that poor households had little option but to place more people into paid work as part of collective survival strategies. Second, the state shifted its rhetoric around PAIT in a way that legitimated the participation of women 'as mothers' in the programme.

Once it became apparent that more women than men had enrolled in PAIT, the state changed its rhetoric around the programme away from one of productive work to one of welfare. Elsewhere, I have discussed the three-stage nature of this shift in detail (Laurie, 1997b). However, it is important to understand the context in which this shift took place and particularly the reasons why the state needed to justify the participation of women in PAIT. Not only had PAIT been designed for men, but it also had a high profile in García's campaign-winning economic reactivation programme. Curbing debt repayments and the radical job creation programme for unemployed men formed the key elements of García's much publicised economic policy. When women rather than men enrolled in PAIT the APRA government had to explain away the feminisation of its flagship programme. APRA did this by shifting its representation of the programme away from the actual work that was done to a focus on the public sector, welfare institutions that PAIT worked with. Consequently, the programme was no longer represented by the state as masculine, productive work but rather as a welfare programme that linked women to public sector, welfare institutions. These women were primarily seen as 'poor mothers' struggling for their families, as the following quotation from an interview with the director of PAIT in south Lima indicates:

> As a programme, we had an experience in PAIT that we weren't expecting. We expected to attract a large number of men as they were supposedly the un-employed in the country. Then we realised that in reality it wasn't like that, because in fact it was women with families who threw themselves into the work-force to struggle for an income, for a salary.
>
> (David Vargas, PAIT Director, south Lima, 1985–1990)

The labelling of the programme as 'welfare for women' was supported by the provision of childcare facilities in PAIT. This allowed women to bring their children to 'work' and to ensure that there would be no tensions between their roles as mothers and workers. The emphasis of PAIT was therefore shifted from work to welfare because women became overly represented in the programme. This welfare representation further alienated men from PAIT, even after the economic crisis worsened in 1987 and employment opportunities for low-income men in formal or informal jobs declined.

What is interesting about this shift in rhetoric, however, is that the nature of the work in PAIT did not change. It continued to be hard, physical labour carried out in public spaces. As we have seen above, it also continued to be work that the women in the programme considered to be 'men's work'. Therefore a contradictory space was created by this shift in rhetoric because women's participation in productive work was maintained and sanctioned by the lie that they were 'merely' welfare recipients within a paternalistic state system. The creation of PAIT as this type of space enabled women to fulfil welfare-mothering roles while holding welfare/production and mother/worker dichotomies in tension. It also helped avoid conflict over constructions of 'good' and 'bad' women. Women's fragmented identities in PAIT were constructed as continua through space and were maintained by making reference to spatial constructs and boundaries. The spatiality of women's engagement in PAIT under these circumstances also changed how women's spaces and femininities were understood.

New locals and new communities

As mentioned above, women's personal isolation was broken down in two ways by PAIT: firstly, as they found friendship and formed new social relations with other women working in the programme and, secondly, as they began to use public spaces more frequently. These changes were experienced in relation to new understandings of geographical scale and notions of community. From a list of nine options covering the three things they enjoyed most about PAIT, 81 per cent of women in a survey of 185 workers said that one of the three things they enjoyed most was the friendships they made; 40 per cent said that one of the three things they enjoyed most was doing something for their community. Therefore, these data suggest that PAIT women defined 'community' in two ways – in terms of a shifting geography of what constitutes the local, and in relation to social communities of solidarity.

The social networks established in the course of working in PAIT reflected the geographical nature of the programme. As PAIT recruited its work-force from within the communities where projects were implemented, many women got to know neighbours who were also enrolled in the programme. New friendships were made between women who had not previously come into contact with each other or who had known each other only superficially prior to PAIT. While the section above mentioned the importance of new social spaces for women, it is also important to note that these new social spaces had particular geographies and territories:

> It was only while working in PAIT that we make friendships with various women from José Galvez. We were from the same place but we didn't know each other so we made friends there. We chatted and sometimes we fell about laughing and talking. (Woman labourer, José Galvez)

The new understanding that women developed between each other in PAIT both stretched the boundaries of the 'local community' and promoted cohesion within a single, bounded, named settlement such as José Galvez:

> The atmosphere was nice because we worked in groups and talked with people from outside of José Galvez and with older people. We would tell each other our problems by asking 'why are you working here?'
>
> *Did you make new friends?*
> Yes, I got to know a lot of women there who were really good. Older women who gave us advice. Me and my friend were younger than the others. Others were our neighbours. We waited for each other, I would let them know about the enrolment: 'Hey, PAIT's enrolling, shall we go and register?' 'Yes, OK then,' so we became united and we would go together. We'd find out where people were working and from José Galvez one group would go somewhere and from Villa el Salvador or Nueva Esperanza they would do the same and we would all go together so that we wouldn't feel alone. (Woman labourer, José Galvez)

The breakdown in the bounding of local space reflected the identification of a common community, a community of women:

What did you like most about the work?
Mostly I liked the *compañeras* [female workmates] because you get used to being there every day, talking. It was fun. You made friends and I got used to that. We would talk about everything and forget about our problems. You forget your problems there because it's a distraction and even though it's with previously unknown people at least you have someone to talk to. You got to know more people and some you got to know very well, all from different places.

(Woman labourer, Villa María)

In Lima and in provincial towns such as Andahuaylas this notion of 'community' was based on the identification of a group of women with the same need, and the formation of an alliance across the differences between women:

Apart from economic things what else did you like about PAIT?
I think that more than anything it was being able to get together with people in the same economic conditions as yourself. I'm not talking about social status because there were students with university education, housewives and illiterate women, there were all sorts, girls with secondary school education and older *señoras* who had completed primary school or some who possibly hadn't. But the economic need was a common denominator, so therefore we shared in the work-place. We chatted, we worked and we helped each other.

(Woman labourer, Andahuaylas)

These alliances worked to undermine a series of differences in the work-place, including class, education level and life course. However, rather than exploding the boundaries around local space so that 'the local' became 'the global', the coming together of the two definitions of community in PAIT shifted understandings of local space as one 'local' became replaced by another larger 'local' community. For example, for Sra Larco, doing something for 'the community' meant improving what she saw as other people's communities, as well as refurbishing the local school where she herself had studied:

Well in my case since we started working we've built tracks and roads, we worked like men to improve things. Obviously it wasn't my area but at least we left something that was well done there and now it's the same here. I'm working in the school where I studied and that's why I'm pleased about it. If they tell me to paint 20 chairs I would finish them by lunch time because as I said to the man in charge 'Because it's my school I'm going to work like a mule', he just laughed and said, 'OK do it!'   (Sra Larco, female labourer, La Tablada)

PAIT therefore provided an opportunity for women to recognise 'their communities' in new ways and to receive payment for activities that they saw as beneficial to others like them. These changes were necessitated by the 'impact' of national and international economic shifts and yet such shifts were filtered through women's 'local' responses to crises where they reinterpreted and expanded their understandings of the local so that 'the global' became 'the local' rather than vice versa.

Shifting the ground of public/private and home and work

Women's experiences of PAIT also raise some interesting issues relating to the spatial dichotomies of public/private and home/work. The first point to make is that, as discussed above, women's involvement in PAIT meant that they not only claimed the work side of the home–work dichotomy but also claimed public (as opposed to private) space. By entering the PAIT workforce, women broke down the gendering of work as they engaged in activities that were previously considered to be the realm of men. They worked beyond the home and showed that women could do paid as well as domestic work. Their paid work relied upon them learning to use 'men's tools' and doing 'masculine work' in public areas:

> The men taught us how to pave the street. At the moment when we were placing the rocks they would show us how to locate them properly, to align them using a level, so that you didn't end up with one sticking out and one sinking down. So sometimes we would take the level and they would put the rocks down. That's the way we learned how to do it.   (Sra Ferafina, Andahuaylas)

Such experiences questioned the distinction between 'men's work' and 'women's work'. Interestingly, however, survey data also suggest that women's new use of public space in PAIT extended to other areas of their lives. For example, as Figure 4.2 illustrates, analysis of changes in decision-making governing the gendered uses of space among these women and their male partners indicated slight shifts in the ways in which women perceived the freedom of both genders to 'come and go'. Specifically, results showed that once women had enrolled in PAIT there was more negotiation between partners over issues of whether women were able to go out on their own (*salir a pasear*) and over decisions about the time until which both partners could be 'out in the street' alone.

These findings therefore indicate that not only did women's direct involvement with the 'public sphere' increase via their entry into the work-place, but also that attitudes towards gendered uses of public space in broader social contexts changed as a result of women's involvement in PAIT:

> Now I go out and I say 'ciao' but it wasn't always like that before. Before we would end up fighting and he would tell me off but I went anyway. But now it's not like that, he just says 'ciao' and 'have a good time' and makes a joke about me meeting someone else.   (Woman labourer, José Galvez)

The second issue of interest is that women reclaimed space through their PAIT activities by evaluating seemingly 'traditional' mothering roles in new ways. This was particularly true in the provincial areas where PAIT operated. Here most women spent the majority of their wages fulfilling household reproductive roles – buying food for the household and providing items for the children. While women labourers spent their money on food, the food they bought was of a different quality or type from that which they had previously bought. Hence, in a survey of 92 PAIT women in Andahuaylas 76 per cent claimed that their household diet improved as a result of their enrolment in PAIT. This is explained by the fact that when PAIT women in Andahuaylas received their wage, many bought certain items in bulk – sacks of rice or sugar and drums

Changes in decision-making concerning whether women with male partners can go out on their own (*salir a pasear*) (as a % of women in the survey)

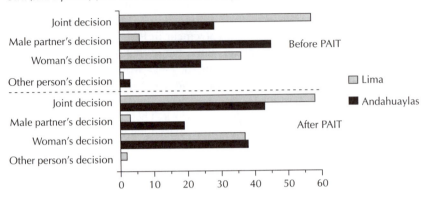

Changes in decision-making concerning the time until when women with male partners can stay out alone (*estar en la calle*) (as a % of all women with partners)

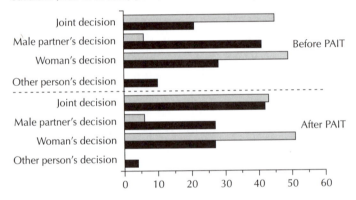

Changes in decision-making concerning the time until when men with women partners can stay out alone (*estar en la calle*) (as a % of all women with partners)

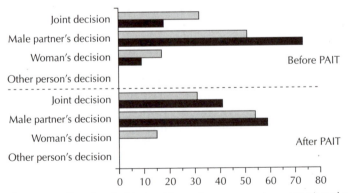

Figure 4.2   Changing decision-making patterns concerning women's and men's use of public space

of oil. One woman explained how the women labourers encouraged each other to spend their money wisely:

> Were we going to fritter it away on momentary pleasures or were we going to buy something important for our homes? Someone recommended that this money, that we were earning with so much effort and sacrifice shouldn't be badly spent. 'Buy something good' she said, 'for example buy sugar, rice, buy specific things.'
> (Woman labourer, Andahuaylas)

With a history of many goods unexpectedly becoming scarce these bulk-buying practices enabled women to budget in a different way. As these goods were imported from the coast they were expensive and consequently items such as rice, cooking oil, sugar and tinned food represented not only basic food stuffs but, it could be argued, they also took on the status of 'high order goods', goods that were rare and costly. These purchases gave the women who bought them a new status even though they were fulfilling their usual reproductive roles in the household by buying food.

These bulk-buying practices in Andahuaylas also prompted women to form new social relations which in turn changed existing individual ways of purchasing goods. For example, in order to buy imported goods in bulk many women in Andahuaylas pooled their wages together to raise capital. The items they bought collectively were then either resold in smaller quantities at a profit or used gradually on a daily basis in individual households. In this way women in Andahuaylas were able to ease the strain of their roles within domestic reproduction in a new way by sharing costs and, as the following quote suggests, by sharing also the physical act of going shopping:

> We made up a group of us and we all went to the shop together to buy our things. We bought the things that we lacked in our houses and in that way the money wasn't badly spent.
> (Woman labourer, Andahuaylas)

As Andahuaylas is an ethnically mixed area with a long history of female illiteracy among Quechua-speaking women, shopping as a group in this way was not just a social outing for these women, nor was it merely a co-operative savings venture. For some women, especially those for whom Spanish was not a first language or those who were innumerate or illiterate, communal shopping was also a learning experience. Sra Ferafina, who had previously worked in an adult literacy programme, explained how important the corporate shopping experience could be for certain women:

> Some of the women didn't know how to work out a budget, or how to add up or work out their money so we used to teach them. We would say, 'Look, with what I've got here I'm going to buy this' and in that way we explained it, focusing on what was lacking in our houses. So a lot of the time we would get them going and say, 'Come on, we're all going to go and buy things together.'
> (Sra Ferafina, woman labourer, Andahuaylas)

Co-operative ventures such as those described above served to cross ethnic and language boundaries and further to cement the relationship among women as they pooled their financial resources as well as their skills in new ways (see Laurie, 1999, for a more detailed examination of changing ethnic identities and the negotiation of racism in PAIT).

In summary, it could be argued that women's roles in budgeting became hybrid because although they still fulfilled their domestic responsibilities by buying food, the meanings attached to that food and the social relations involved in budgeting were different. Concentrating only on the fact that women spent wages on seemingly traditional female purchases such as food obscures the changing nature of the social relations around those purchases. Such an approach also ignores the symbolic value of the exchange and any resulting changes in women's negotiating position within their households.

A third observation on the women's experiences in relation to spatial dichotomies is that, while women's involvement in PAIT allowed the work and public sides of the dichotomies to be claimed by women, often for the first time, in conceptual terms the boundary between the two sides of each of these dichotomies also became blurred by women's participation in the programme. The most important blurring occurred as women in PAIT identified solidarity with their fellow female workers and spoke of the companionship they found in the programme. This companionship enabled them to talk freely, which was something they were only used to doing within the context of the family. Consequently, they were able to air their personal, 'private' problems in 'public' and find that they were not alone in their experiences:

> We got experience there because we learnt about different problems. Everyone has problems and we talked about our problems and we gave each other advice. That's why I worked almost every time in PAIT. (Woman labourer, La Tablada)

Younger women, in particular, mentioned the importance of having an opportunity to hear the experiences of older women whom they got to know. While these older women became friends they were not family members and so their advice could be heard by the younger women without resentment or accusations of interference:

> *What did you like most?*
> Some of the older mothers helped the younger ones to understand things: 'That's how you should behave like that, like this.' (Woman labourer, La Tablada)

In this way the distinction between home and work spaces and public and private information became blurred. This blurring was also evident in the fact that many women took food they had prepared to the work-place and shared it:

> I had a neighbour who worked with me. We would take things together. Sometimes we would take food jointly and eat together. She would take one thing and I would take another and we would share it out there. We would also travel together. Sometimes I only went with my baby and left the others behind so I could help her carry her kids. Also, we would go together to the bank. One would go first to get in the queue and save a place for when the other one arrived.
> (Woman labourer, Villa María)

In PAIT there was a two-way shift between the public and private: by sharing food women took women's solidarity and 'private' activities into the public sphere, and by sharing personal problems with strangers (non-family members) 'the public' were taken into 'private' worlds. This blurring of the public/private boundary meant that new friendship had a long-lasting influence on the ways in which women related to each other:

> *Did you make friends?*
> Yes, I made a lot of friendships. I made many, many friends and ever since then when those of us who worked in the programme see each other we always have a unique bond of kindness. Why, you ask? Well because we are all women who shared a job, we struggled together. So therefore we are always glad to see each other. When we meet we greet each other very fondly.
>
> (Woman labourer, Andahuaylas)

As advice was sought and given, fellow workers became 'family':

> They were like family, like my sisters.   (Woman labourer, Villa María)

> We formed what was almost a sisterhood, to such an extent that if we didn't meet or if one day someone didn't go we missed each other.
>
> (Woman labourer, La Tablada)

Therefore, the breakdown in the distinction between public and private space was also coupled with a blurring of the boundary between home and work as fellow workers became 'adopted family'. The home/work boundary was also blurred by the fact that women claimed that they had fun 'at work':

> It was a bit of fun between friends, the time went quickly. You would forget your sadness [*pena*] when you were upset or something, talking would cheer you up. Some gossiped but when some of them had a problem at home, something like that, you would forget it, you'd distract yourself – have a bit of fun – working.
>
> (Woman labourer, La Tablada)

Several women also used the skills they had learned in the work-place to improve their own homes:

> *And have these things [skills learnt at work] been useful to you?*
> Oh yes, a lot, for example all that about paving, afterwards I paved my own patio in the same way so that the house didn't always get full of mud.
>
> (Sra Ferafina, woman labourer, Andahuaylas)

In Andahuaylas one group of women took turns painting and plastering each other's houses. One woman said 'I wanted to surprise my husband and when he came back from Lima he couldn't believe we women had plastered the wall' – the house was made of adobe mud brick and she and her friends had plastered it with whitewash cement. Women therefore claimed new ground in PAIT. They took their 'work' into their 'home' and their home into their work. They shared what were normally individual, private, domestic tasks and aired their personal, private issues publicly.

Superseding boundaries

While it is important to show the different roles that dichotomous understandings of space played in the construction of femininities in PAIT, it is also important to note that not all changes are captured by making reference to bounded space. Women's experiences of PAIT also superseded dichotomous ways of conceptualising space. The constitution of femininity in PAIT was dynamic and involved processes that questioned the hierarchies that structured both the work-place and the home. For example, in the face of corruption and hyper-inflation women took to the streets to complain about the failure of the state to pay them:

> Everyone was annoyed, at least everyone who worked in the *campo*. They said 'Why have they said that they are going to pay us on this date when people don't have money now and when they are dying of hunger and are broke? How long is this going to go on without them paying us?' So someone said 'we have to go on a march' and everyone went.
>
> *Did you get what you wanted?*
> Yes, after that they paid us.   (Woman labourer, Villa María)

The strength of women's solidarity experienced by individuals provoked rebellion in the ranks as women formed political alliances across their differences:

> Everyone's different, aren't they? But when they didn't pay us – well they were meant to pay us every 15 days but sometimes it would go for more than a month – some women said 'That's enough, I'm going to shout, you have to say "that's enough".' And that's what we did. There were always some women who led it.
>   (Woman labourer, La Tablada)

In some cases this solidarity led to a questioning of the value of the work itself:

> Sometimes with my *compañeras* we used to say that we couldn't carry very big rocks, we decided to carry small ones especially for the wage we were getting. So we all agreed and only carried small rocks.   (Woman labourer, La Tablada)

While many women spoke of how the public face of solidarity helped change attitudes and enabled them to resist oppression, other women spoke of the strength they gained from this solidarity for their own private battles over the mother–worker tension. For example, this solidarity helped one woman, Sra Martínez, to continue arguing with her male partner about her 'right to work' once her involvement in PAIT stopped:

> I'm sure PAIT helped me to have my own opinions. It helped in overcoming the prejudices of my husband. For example now I'm a teacher and I go long distances, I work away from here. Now he's not jealous, we don't fight about it.
>   (Woman labourer, Andahuaylas)

These changes therefore out-lasted the PAIT programme itself. Some authors such as Rowlands (1997) and Kabeer (1994) would suggest that this type of questioning constitutes empowerment (see Chapter 2).

Finally, the experience of women in PAIT shows that changes in femininities occur in a continuum and in relation to other experiences. For example, women who had been involved in community projects such as soup kitchens before PAIT learned to think differently about themselves and about the things they were doing in these groups. The case of Sra Martínez, the leader of a soup kitchen funded by the charity, World Vision, in José Galvez illustrates this point well. I quote this exchange in full because it shows how important it is to recognise that the relationships between seemingly discrete personal and public experiences/changes are central to the construction of femininity:

> Sometimes [in the club] I used to be afraid of the other mothers, because of their criticisms. . . . In PAIT I learned a lot about how to be more free to answer back and that not all [women] are the same. Sometimes when we arrived late the supervisor would say that they were going to dock our pay. Some mothers complained and they said to me, 'Why don't you protest too, shout about it, you have kids too?' Until then I had always been quiet.
>
> *Was this the first time you said such things?*
> Yes.
>
> *Did these experiences influence you in the club?*
> Without PAIT I couldn't have been vice president of the club, I learned a lot in PAIT. I even have a neighbour who comments on it. She says 'since you went into PAIT you've changed a lot. You were too quiet, now on some things you answer back'.
>
> *Has that changed your relationship with your kids, with your husband?*
> Well probably with my husband because before when I didn't answer him back he used to beat me and I just stayed quiet. But not now, now I answer back and he doesn't hit me.
>
> *Does he say anything about the club?*
> Yes, he says 'what are you doing involved in that? It's not as if you are earning there, so what do you do?' And so now I always tell him about the club.
>
> (Sra Martínez, woman labourer, José Galvez)

Sra Martínez's experiences in PAIT helped shape her experiences of 'community'. In turn these experiences helped shift the gender politics in her household. The relationship between these three sites of change is transformatory and cannot be captured by geographical interpretations that fix women's experiences in dichotomous understandings of space. Instead such changes can only be understood in the context of the spatiality of women's work in PAIT, whereby space and spatial understandings both shape and are shaped by the constitution of femininities of low-income Peruvian women.

Conclusion

By focusing on state-led emergency employment as a site for changing femininities this chapter has attempted to show how both national and international restructuring and crises help shape the contexts for the negotiation of new femininities. It has illustrated how the spaces produced such by changes are reconfigured both by changing rhetorics of welfare and by women's own actions in representing themselves and their relationship with their local communities and the people who constitute them. Central to the actions of individual women is the contestation of notions of 'appropriate' gendered behaviour. These are challenged and expressed through the everyday re-evaluation of existing gender norms and by complex geographies of representation. The chapter points to the fact that these geographies are at the heart of the negotiation of new femininities as they allow fractured gender identities to be maintained and reworked in complex and changing ways.

CHAPTER FIVE

Reproducing motherhood

SARAH L. HOLLOWAY

Introduction

Motherhood is often regarded as a natural phenomenon: it is looked upon as normal for a woman to want to bear and raise children, and in so doing she is seen to be fulfilling what it means to be a woman, and hence to be fulfilled as a woman. Indeed, motherhood is a central feature of many women's lives all over the world, and it unquestionably differentiates their experiences from those of men. A book on femininities then, would not be complete without mention of motherhood. However, it is clear on closer inspection that although motherhood is of fundamental importance, it is not, in any simple sense, a natural phenomenon. For one thing not all women do want to have children, and the desire of some others to bear children is regarded as selfish rather than normal, as the women in question are deemed by their wider society to be too old, too young, to have too many other children, or to be from an unsuitable minority group, etc. (Phoenix and Woollett, 1991a, 1991b; Bartlett, 1994; Letherby, 1994). Moreover, the ways in which mothers raise their children differ radically both within and between societies, and over time (Richardson, 1993; Newson and Newson, 1974). As Dally (1982: 17) argues:

> There have always been mothers but motherhood was invented. Each subsequent age and society has defined it in its own terms and imposed its own restrictions and expectations on mothers. Thus motherhood has not always seemed or been the same.

The differences in contemporary British society between the terms mothering, which implies a caring role, and fathering, which connotes a biological role in conception, further illustrate how clearly gendered are our ideas of parenthood. Far from being a simply natural experience, motherhood is a complex social phenomenon: it varies over time and space, and is intimately bound up with normative ideas about femininity.

This chapter, like the previous one, provides a case-study examination of the ways in which femininities are being reproduced, in this case by focusing on mothering practices in contemporary Britain. Britain, like Peru, has been changed by developments in the world economy over recent decades, and has also experienced home-grown social and political transformations. These changes mean that British mothers are currently raising their children in contexts different from those of the previous generation, and as a result mothering practices are being renegotiated. This chapter explores how different groups of mothers, in particular 'white', heterosexual mothers from middle- and working-class backgrounds, are negotiating motherhood and childrearing in this new context. Specifically, it concentrates on two areas where the reproduction of ideas about 'good mothering' can most clearly be seen: first it analyses the reasons why some mothers care for their children full time and others undertake full- or part-time employment, and second it examines how 'working' and 'non-working' mothers alike ensure that their children gain access to a preschool education. The final section of the chapter then turns to an analysis of the importance of geographical practices in the reproduction of mothering, before conclusions are drawn.

The changing context of mothering in Britain

The world economy, and Britain's position within it, has changed rapidly in the post-war years. After World War II manufacturing in Britain was buoyed by emerging consumer markets, but by the 1970s problems were apparent, and the manufacturing sector and, in particular, older industries such as cotton and steel production suffered accelerating rates of decline through the 1980s. The devastating losses of employment were only partially offset by a growth in the service sector, which for some regions and some workers provided new opportunities. Thus some men who had worked in traditional industries found themselves 'unemployable', while part-time, poorly paid 'jobs for women' and high-quality service sector jobs for some professionals grew in number. The changes, combined with a long period of Conservative government from 1979 to 1997, meant that the gap between the rich and poor in Britain, which had been declining in the post-war period, started to grow again (see Champion and Townsend, 1990, for more details).

These profound economic changes occurred alongside widespread social change. One significant change has been the growth of employment among women, and of particular interest to this study, among women with young children. The General Household Survey shows that labour-force participation by women with pre-school aged children increased from 25 to 43 per cent in Britain between 1973 and 1991 (OPCS, 1993) and this figure appears to be rising through the 1990s (ONS, 1998). This change in social practice has been accompanied by changes in attitudes with fewer men and women from the 1950–1970 birth cohort thinking that a women should stay at home when there is a child under school age (men 48 per cent; women 46 per cent) than is the case for those born between 1930 and 1950 (men 76 per cent; women 70 per cent) (Scott *et al.*, 1993). The result is that fewer women care for their young children in the home than was the case a generation ago. Some of these working mothers have careers with good salaries and good employment prospects;

however, many others work in low-paid jobs in order to supplement household incomes. The post-war period in Britain has also seen transformations in
household form through legal and attitudinal changes to divorce (current
divorce rates mean that roughly two in five marriages will end in divorce) and
birth outside marriage (9 per cent of all births in 1975, 34 per cent in 1995)
(Smith et al., 1993; ONS, 1998). One consequence of these changes has been a
rise in the proportion of households with dependent children that are headed
by lone parents, mainly lone mothers (ONS, 1998). Such changes have been
one stimulus to debates about the role of fathers in the 1990s (Moss, 1995),
especially in relation to absent fathers; however, although social policy has
reinforced fathers' financial responsibility for their children (Fox Harding,
1994; Millar, 1994), childcare is still constructed as women's responsibility
(Marshall, 1991).

These socio-economic changes have brought about remarkably few changes
in government policy towards the care of children with working parents.
During the two World Wars the government provided free nursery provision
as part of a package of measures designed to enable women to participate in
paid work, but in peace time these services were cutback and the provision of
care for children while their mothers worked was redefined as a private
responsibility (Dale and Foster, 1986). This definition of childcare as a private
responsibility has remained entrenched throughout the 1980s and 1990s
(Fincher, 1996), and as a result working mothers in Britain are dependent on
a mixed economy provision (OPCS, 1993). Commercial care has long been available from childminders and in the recent past there has been a resurgence in
nanny employment and an expansion in private nursery provision; however,
many working mothers continue to rely on other family members, principally
their own mothers, for the provision of care (Cohen, 1988; Gregson and Lowe,
1994; Meltzer, 1994).

In contrast, successive post-war governments have held positive attitudes
towards publicly funded pre-school education, though spending has been concentrated on the needs of older children and young people, and responsibility
devolved to the local level. Some Local Education Authorities (which until April
1997 were responsible for publicly financed nursery education) strove to provide universal access to free, part-time nursery education for 4-year-olds in their
area, but in other authorities no state nursery education places were funded
(Cohen, 1988; Owen and Moss, 1989). The Playgroup Movement, which
emerged in the 1960s with the initial aim of providing temporary provision
in areas lacking state nursery education, was quickly successful, especially in
middle-class suburban areas where mothers became involved in providing the
part-time care where children could learn through play (Finch, 1985; Brophy
et al., 1992). The Pre-school Learning Alliance now promotes playgroups, recently renamed as pre-schools, as a long-term alternative to nursery education.
Private nurseries and nursery schools are also providing an increasing amount
of educational care for the children of working and non-working parents alike,
though the cost to parents of this form of provision is higher than that for state
nursery education or playgroups. In the mid-1990s nursery education worked

its way to the top of all parties' political agenda, and since the fieldwork for this study was undertaken, new funding mechanisms have been introduced.*

The Sheffield case study

It is in this socio-economic and policy context that mothering practices are being reproduced in contemporary Britain. This chapter explores the reproduction of these practices, for particular groups of mothers, through case-study research in two contrasting areas of Sheffield, one of England's northern cities (see Chapter 3 for details of the methodology). Sheffield has a long history of cutlery production and grew rapidly in the nineteenth century with the expansion of 'heavy' steel production. The dominance of steel in the local economy brought both an international reputation and wealth to the city. The concentration of unionised steel workers also stimulated the development of a powerful working-class-based political movement, and the Labour Party has been in near continuous control of the City Council since the mid-1920s. Despite the historical importance of women in the labour market, for example as 'buffer girls' in the cutlery trades, women played only a minor role in these political movements (cf. Mark-Lawson *et al.*, 1985). The economic restructuring of recent decades, and the consequent decline in the importance of the British steel industry, has, however, left Sheffield with a tarnished image as a declining industrial city and considerable economic problems, particularly in terms of unemployment (Watts *et al.*, 1989; Taylor *et al.*, 1996).

The city has also been left with a spatially polarised social structure. One case-study area, Hallam, is located in the more affluent south-western sector of the city, and contains within its boundaries some of the more desirable addresses in Sheffield. Southey Green is a working-class area, a 15 minute drive north of the city centre, comprising principally low-rise, low-density estates where many residents rent their homes from the city council (see Table 5.1). A questionnaire survey of households with pre-school aged children in Hallam and Southey Green highlights considerable differences in their socio-economic position (see Table 5.2). Their different class position is evident in both the employment histories of primary childcarers, who tend to have held higher status jobs in Hallam, and current household incomes, which are much lower in Southey Green. Unemployment among secondary childcarers (mainly fathers)

---

* The funding of 'educational' care for 4-year-olds has been through a period of rapid change since the fieldwork for this study was undertaken. In April 1996 the Conservative Government started to pilot a nursery voucher scheme, and this was introduced nation-wide from 1 April 1997. Nursery vouchers were issued to parents of 4-year-old children; these could be used to pay for a part-time nursery or playgroup place in the state, private or voluntary sector, if provision was available. In May 1997 the newly elected Labour Government abolished the nursery voucher scheme, which it argued reduced choice, increased bureaucracy and created very few new places. From September 1998 the Government has promised to provide a good quality pre-school education for all 4-year-olds whose parents want it. Provision is now co-ordinated at the local authority level, and depends upon partnership between the public, private and voluntary sectors. It is too early to assess the outcomes of this policy.

Table 5.1 Socio-economic statistics for Hallam and Southey Green, 1991 Census

	Hallam	Southey Green
Total population	17,693	15,512
% of population who are white	97.1	98.6
Class structure	69.8% I and II	77.1% IIIM, IV and V
Housing tenure	84.7% owner occupied	73.7% council homes
Car ownership	75.9% one or more cars	62.2% no car
Unemployment	4.7%	21.4%

The city-wide unemployment rate is 12.4%.
Source: CPU (1993)

Table 5.2 Socio-economic composition of households with pre-school aged children in Hallam and Southey Green*

	Hallam	Southey Green
Measures of class background		
Class position of primary childcarer before birth of first child	45% professional or managerial, 26% skilled non-manual	48% in manual work (mostly semi-skilled), 26% not in paid work
Income	43% net income > £20,000 p.a.	43% net income < £5,000 p.a.
Other measures of social background		
Current employment among secondary childcarers	96% full-time, 1% part-time, 3% not in paid work	57% full-time, 2% part-time, 41% not in paid work†
Current employment among primary childcarers	17% full-time, 29% part-time, 54% not in paid work	3% full-time, 12% part-time, 85% not in paid work
Ethnic origin	95% of parents 'white'	97% of parents 'white'
Household structure	99% of households consist of heterosexual nuclear families, where the woman takes primary responsibility for childcare in 96% of cases	69% of households consist of heterosexual nuclear families, where the woman takes primary responsibility for childcare in 98% of cases. In addition, 31% of households are headed by lone parents, 90% of whom are women

* Data obtained by questionnaire survey.
† The figure of 41% may be inflated by under-reporting of economic activity; however, the 1991 census shows that unemployment is a major problem in this ward, with an average rate for the ward as a whole being 21.4%.

is also much more of a problem in Southey Green. Other social differences between the areas are evident in the higher rate of employment amongst primary childcarers in Hallam, and the broader mix of one- and two-parent families in Southey Green. However, two striking similarities also emerge. The vast majority of residents in both areas are 'white', and, importantly, though not surprisingly, the questionnaire survey shows that it is overwhelmingly mothers, not fathers, who have primary responsibility for the care of children in both Hallam and Southey Green. It is to an analysis of these women's mothering practices that the chapter now turns.

Mapping mothering

Local childcare cultures

Women's mothering practices, it is argued here, are shaped within the context of a local childcare culture. Two examples, one of which focuses on motherhood and paid employment, the other on the development of the pre-school child, are used below to demonstrate the ways in which these local childcare cultures are reproduced through the differential form of, and interaction between, mothers' attitudes towards childrearing, their cultural and material resources, and the childcare provision available. Mothers' attitudes to childrearing develop out of their own experience, information gained through their social networks, and information passed on through childcare professionals and other media. Their cultural and material resources, which stem from their social status and their social networks, enable and constrain mothers in choosing how to organise their everyday lives and care for their children. The amount, form and function of childcare provision available to mothers are influenced by childrearing ideologies, national and local party politics, the position of the areas in wider socio-economic structures as well as the activities of other groups and individuals. The importance of geographies and the emergence of new femininities in these local childcare cultures is drawn out in the two examples and considered in more detail later in the chapter.

Motherhood and paid employment?

The increases in the employment of women with young children described earlier have not been evenly spread in the two case study areas. In Hallam 17 per cent of primary childcarers work full time, 29 per cent work part time and 54 per cent are not in paid employment. The employment histories of mothers in this area show a considerable level of stability: many mothers who work have been in paid employment throughout their childbearing years, and many who do not currently work left their job when they had their first child. In Southey Green fewer primary childcarers are in paid employment. However, while only 3 per cent of primary childcarers currently work full time, and a further 12 per cent part time, the employment histories of the mothers interviewed here reveals that many who are not currently active in the labour market had been in paid work at some point since the birth of their first child. In contrast to Hallam, the employment situation of these mothers was relatively fluid and many were moving into and out of short-term, usually part-time employment. The reasons for these differences in employment patterns are complex, and include mothers' own attitudes to motherhood, paid work and childcare, their cultural and material resources (particularly their position in local labour markets) and provision availability. The interaction of these influences on mothers' employment status is examined below (but see Holloway, 1999, for a fuller examination of the issues discussed in this example).

Attitudes to motherhood and paid employment

Some mothers in both Hallam and Southey Green want to work because it is important to their own well-being. Friedon and Gavron, writing in 1963 and 1966 respectively, played an important part in early feminist attempts to highlight the problems of isolation experienced by women raising their children at home. The interviews in Hallam and Southey Green show that isolation is an ongoing problem for some, and is one reason why some mothers choose to return to work:

> When I had Heather I think I was a little bit depressed, I was very isolated so I think I had to go back to work because I felt like I was going off my head, I really did.   (Fiona, working mother, Hallam)

> I found it very depressing at first. I'd worked from leaving school, I'd never been, well I had been unemployed but even then I was on a job training scheme, so I have never done nothing, and I had no money, no family close . . . and it was awful.   (Suzy, working mother, Southey Green)

Isolation, however, takes a diversity of forms and while some women had narrow social networks, others had family and friends living locally to whom they could turn but found themselves isolated from non-mothering networks. In particular, some mothers felt isolated from the self-esteem work-place interaction can bring in a society where your job is regarded as an integral aspect of your identity. In the quotations below Vanessa and Sophie explain how important working is to their sense of self-confidence and self-worth:

> I think it does you good to go back to work, part-time anyhow, I lost my confidence a lot. I mean even to the point of picking the phone up and having to make a phone call . . . but it is coming back now, I'm enjoying myself [at work] and it gives me a break.   (Vanessa, working mother, Southey Green)

> I've got to be me, and this is really why I am looking for a job now, because I am not me any more, I am just somebody's other half, be it mother or spouse, and somewhere I am getting lost and I need, I need somebody to say, 'yes, you've done a good job there' because nobody does that when you are a mother. They tell you that your child is badly behaved and you are pathetic, well they don't [because mine aren't], but nobody says, nobody comes up and says, 'wow' once a year, 'you've done a really good job there, keep up the good work and we will give you a pay increase!'   (Sophie, previously working mother, Hallam)

The sense of independence gained through working is also reinforced by the money women earn. In Hallam some women's partners earned enough for them to construct their earnings as a secondary benefit of working, effectively characterising their wages as pin money which paid for the extras after their partners had provided for the family's basic needs. However, in Southey Green, though the mothers might earn less than those in Hallam, earnings were seen as an important contribution to the household budget, raising the family's standard of living, and as particularly important in giving women money they could call their own.

These mothers, and others, also placed a great emphasis on the importance of the role mothers play in the home. Mothers see themselves as crucial in providing their children with love and with a sense of security and safety, while at the same time shaping their children's fast-developing personalities. These beliefs are widely held, but for some mothers the desire to undertake this role full time is so strong that they rule out paid employment outside the home:

> I mean I don't condemn people that do go out to work, but I think, you know, a mother should be there for her children . . . I am a firm believer that if you put in now you reap the benefits [later]
>
> (Lydia, mother not in paid work, Hallam)

> I feel like I want to be a mum really, I want to make the most of it until they get, like Catherine's age when they are at school.
>
> (Lorraine, mother not in paid work, Hallam)

> I mean, I'll go back to men with their chauvinistic attitude, oh women should always be doing this and that, and in the house and everything, but to be honest I agree with that to a certain extent. I wouldn't like it if there were mothers, well I mean it is up to them, but I wouldn't want to be that type of person that would just leave them [my children] to it.
>
> (Sally, mother not in paid work, Southey Green)

These women have made a positive choice to work as full-time mothers; others, however, rework these understandings of maternal responsibility in a way that makes them applicable to a social world where paid employment is part of mothers' everyday lives.

This reworking of what we might call 'traditional' visions of maternal responsibility takes a different form in Hallam and Southey Green. In Hallam working mothers make up for not being there all the time by focusing on child-centred activities (which would often be too expensive for parents in Southey Green) outside working hours and/or by working part time:

> Our typical weekend is very child centred. Saturday morning they [children aged 4 and 8] both have tennis lessons . . . so the whole Saturday morning is devoted to tennis and we go to the library via the way home and then, ur, what did we do this Saturday? We went for a bike ride this Saturday, in the Peak District somewhere, and then Sunday we go swimming and then Sunday afternoon we went to the Merlin Theatre because they are doing a lot of children's theatre things,

puppet shows, mime and that sort of thing, so we went there. . . . I think obviously with working you don't see them in the day so you do actually have to
make good use of your time, that you've got, we've always believed that, that
you make up, you can make up for not being there during the week. And it is
the same in the holidays as well . . .   (Dilys, working mother, Hallam)

In Southey Green part-time working comes to be seen as one way of fulfilling
maternal responsibility as women are seen to be working for their children,
that is working for money to improve their children's standard of living:

> I mean they are going out to work for their children, aren't they? To better them-
> selves so they have got more for their kids.
>   (Clare, previously working mother, Southey Green)

Thus mothers in both areas are actively reworking the concept of maternal
responsibility to apply to the specific circumstances of their lives. This does
not mean that 'traditional' understandings of a mother's role in the home are
erased, rather that new understandings of 'good mothering' are shaped and
defined in relation to this. The previously dominant ideology is not necessarily transformed but its reproduction is blurred.

Mothers' cultural and material resources: the importance of
labour market position
However, while mothers' attitudes to childrearing and paid employment are
important, it is not simply the case that more mothers favour working in Hallam,
and thus that employment rates are higher here. Other factors, notably mothers'
position in the labour market and the differential availability of, and access
to, childcare provision also come into play. In Hallam many mothers hold or
have held professional jobs, for example as opticians, accountants, teachers and
civil servants, that require long periods of training and are relatively highly
paid. Many fathers also enjoy a relatively privileged position within the labour
market. In Southey Green many mothers hold or have held semi-skilled manual jobs, for example in shops, sweet factories, bakeries and pubs, that pay a
relatively low wage, and a sizeable minority have never worked. Their poorer
economic position is compounded by the fact that roughly a third of households with young children in this areas are headed by a lone parent, and in
those with two parents low wages and unemployment are a problem among
fathers (see Table 5.2).

These differences in labour market position, and the consequent differences
in cultural and material resources, influence mothers' employment and childcare 'choices'. Many women in Hallam have more to lose by not working than
mothers in Southey Green, as their forms of employment tend to have career
structures that are not compatible with taking time off to raise children. If they
opt to take time out to raise children they are likely to suffer what is called
'downward occupational mobility' (see the classic survey of Martin and
Roberts, 1984), that is on returning to paid employment they are unlikely to
be able to secure as good a job as they had previously held. Women for whom

promotion is very important are under considerable pressure to go back to work full time:

> I'm quite an ambitious individual and I wanted to do a lot more at work. There was never ever a chance of not going back to work, it was always 'I'm going back and I am going back full time', which is why we won't have any more [children].   (Lindsey, working mother, Hallam)

Others may opt for part-time work. This allows them to maintain their current position in the labour market, and the official and unofficial benefits that that brings (McRae, 1991). For example, Julia continued to teach part time when she had her children because in the long run she wanted to continue working and she had reached the position where she was teaching the classes she wanted within a school:

> It never really crossed my mind [to leave work] because to get back in is just . . . , even when I was sort of considering that five years ago, it was beginning to be difficult to get back into teaching in Sheffield, so those considerations, also I teach the students I really like, I wouldn't have wanted to start again.
>
>   (Julia, working mother, Hallam)

In contrast, mothers in Southey Green already occupy a position near the bottom of the labour market, and thus moving in and out of paid employment is unlikely to result in downward occupational mobility. As Phoenix (1991) has pointed out in her study of mothers under 20, the conflictual model of motherhood or high-quality employment does not apply to all women. However, it is not only the case that mothers in Hallam have more to lose from not working; they are also more likely to be able to gain from working. Their higher earning potential means that they can afford to use (though not without difficulty) a range of non-parental childcare services. In contrast mothers in Southey Green can become stuck in a poverty trap because they cannot afford to leave state benefits and/or to pay for childcare provision:

> The thing is, at the moment my husband is not on a very good wage, we get family credit [a state benefit]. If I went to work then we wouldn't get, you know, once you get family credit you get free prescriptions and free eye tests and things like that, and then it would be, and we'd probably stop claiming that, so I may as well stop at home. I know it sounds wrong but. . . .
>
>   (Clare, previously working mother, Southey Green)

> Jobs, I still look in paper for jobs. I don't know why I bother because the wages are so low that I wouldn't be able to afford anyone to [baby]sit, and it is a case of finding someone to sit, I'd be working for nothing. . . .
>
>   (Alison, previously working mother, Southey Green)

Childcare provision for working mothers
Access to childcare provision is indeed an important consideration in both Hallam and Southey Green. Some mothers in both areas are able to secure access to non-parental childcare provision despite limited financial resources because other

resources, their social networks, give them access to care provided by other family members. In some cases the care is provided by the mother's partner. For example Julie has found work that fits in with her husband's working hours:

> I only do six and a half hours [work a week] but it's times I want, you see, cos like, you know, with having during the day and that and, like school holidays, I've got to have something what qualifies with John coming home so he can have the kids while I go to work. . . . I've been looking for ages for something with them hours, with him coming home so I don't have to rely on anybody to look after them.   (Julie, working mother, Southey Green)

However, grandmothers are also a very important source of care. This family care is highly valued because it is generally free and because it is seen as a very safe form of provision. The children know their relatives and the relatives are known to the parents who trust them. Moreover, the care takes place in a family home, the environment most often considered suitable for young children (Brannen and Moss, 1988). It does, though, have important drawbacks: if women find work to fit in with their partner's hours they are usually limited to part-time evening work; if grandmothers are relied upon, women tend to work only part time for fear of 'putting upon' (over-using) their relations.

The incomes of those households that use family care are not that dissimilar between the areas. Those households that use family care in Southey Green tend to have higher than average incomes for the area, in part because family care allows the mothers to work. Those households interviewed who use family care in Hallam, especially those who rely solely on family care, have particularly low incomes for the area, meaning other forms of care are beyond their reach. One difference between the areas comes in the attitudes of those who do not use family care. In Southey Green those households who did not use family care said that family circumstances made this impossible; for example, grandparents were too old, worked themselves or did not live locally. Indeed the capricious nature of family care is one reason why so many women in this area move in and out of employment. It is the only form of care to which many mothers can gain access; however, its availability is often temporary. While some mothers in Hallam also explained that family care was not available, a number of others stated that they would not want to use their parents for care because their ideas about childrearing were not compatible. The reason why some mothers in Hallam can choose to reject extended-family provision is that their greater financial resources, highlighted above, provide them with a wider range of childcare choices.

These choices include care by childminders and private nurseries, in addition to other less frequently used services, all of which are out of the financial grasp of most parents in Southey Green. Mothers' attitudes to these different forms of provision, which stem in part from their localised information networks and their own experience of using the services for their children, are important in shaping their choices among these alternative forms of non-familial childcare provision. Some mothers, like those in Dyck's (1996) study, believe that maternal care in the home is best for children and consciously try to reproduce this environment for their children by using childminders while they are out at work. These mothers value the relationship that the child can enjoy with a woman who is likely to be a mother herself, and who will love and care for the child as part of a family-type unit:

> I didn't feel that it [a nursery] was the right environment for a baby, I was feeling fraught about that. . . . I thought she would be better off with a childminder, sort of in a home environment maybe with a couple of other kids.
>
> (Fiona, working mother, Hallam)

Finding the childminder through personal links, such as recommendations from friends and family, enhances the value placed on this form of care:

> I was quite interested [in having Vanessa as a childminder] because Lizzie knew her and that was, and she said that she was very, very good. She was very kind and she was very understanding and that she came highly recommended. And so Matt and I went to see her and we really liked her, she was really nice and Joshua loved her and, you know, she was like a second mummy to him really.
>
> (Ruth, working mother, Hallam)

Many of the mothers who were using private nurseries at the time of interview had previously used childminders. Problems with the quality and continuity of care had caused some mothers to change provision type; however, there is also a sense in which mothers see the move from childminder to nursery as a natural one, enabling older pre-school children to benefit from a stimulating educational environment:

> It seemed a natural progression somehow for him to be with a childminder at the beginning, when he needed, there was that degree of dependency, and then for him to move out, to get used to being with other children and now he is almost in that pre-school position of being one of the elder ones [at his nursery]
>
> (Lindsey, working mother, Hallam)

> [His nursery] is just for 2-year-olds plus and they actually think about preparing them for school, when they, as they get older, I mean Charles doesn't, he is still at the very bottom end of, he has got plenty of time, but they actually, they do pre-school stuff, they do things like colours and shapes and so on and it is a really nice place and its got lots of lovely equipment and we just really liked it. . . .
>
> (Julia, working mother, Hallam)

The availability in Hallam of collective childcare provision of a similar cost to home-based childcare is resulting in the use of a range of childcare 'spaces'. The collective spaces, such as private nurseries, challenge the spatial prescription of post-war childrearing ideology that home care is best. This challenge to the importance of care by one woman in the home is mediated in different ways for children of different ages. Nurseries that provide care for very young pre-school children tend to minimise the challenge to the idea that home care is best by conforming to the ideal of individual care by one adult as closely as is possible within a nursery environment. This emphasis on continuity in the baby–carer relationship, for example by making named carers responsible for the care of individual children, makes mothers feel happier about using nursery care for very young children. In contrast for older pre-school children the positive difference from home care is stressed by both nurseries and mothers, who emphasise instead the benefits of a stimulating nursery environment in the development of the pre-school child. It is to the question of pre-school education that the chapter now turns.

Developing the pre-school child

Mothers in Britain have been held responsible for different aspects of their children's well-being at different times. In the early years of this century the newly emerging childrearing experts started to focus parents' attention on the need for physical and mental hygiene in the raising of children, and away from the more 'traditional' concern for the need to discipline children in order to 'defeat the devil within'. By the middle years of the century the 'hygienic' approach was replaced by the 'permissive' approach, which emphasised the need for mothers to respond lovingly to their children's needs in order to ensure their mental health (Newson and Newson, 1974; Richardson, 1993). Analysis of contemporary childrearing manuals shows that mothers continue to be held accountable for children's mental and physical health (Marshall, 1991), and Richardson (1993) argues that they are also increasingly being held responsible for the educational development of their pre-school children. This section explores how mothers' attitudes to pre-school education are reproduced in the case-study areas, and considers the diverse ways in which they access educational provision for their children (see Holloway, 1998b, for a fuller examination of the issues discussed in this example).

Attitudes to pre-school education

Mothers in both Hallam and Southey Green are certainly keen to ensure that their children receive some form of pre-school education. They want their children to get used to spending time apart from them before they have to go to school five days a week (at the age of 5), and also want them to mix with other children, both to enjoy the company and to get them used to being in a group of children their own age. However, although there is general agreement that pre-school aged children need some form of education to prepare them for school, there are major differences in the ways in which mothers in the two areas talk about this need and how it might best be met. Mothers in Hallam are clearly used to talking about pre-school education and are happy to engage in debate over the relative merits of different pre-school learning environments, which in this area means playgroups, LEA nursery school or private nurseries. For example, in the quotation below Jane explains why she prefers private nursery provision to playgroups:

> [Playgroup] is much more sort of activities based around slapping paint on paper
> and that sort of thing, and all having a sing-song at the end, whereas at [private]

nursery Polly does 'Letterland' which is one of the building bricks of learning how to read, letter recognition and that sort of thing. They spend a lot of time doing sort of pre-reading skills and easy number work so that, all right, by the time they get to school they can't read, they've not been taught to read by nursery, but they are certainly along the road, whereas at playgroup it is very much: there is a nice activity going on at this table, there is a nice activity going on there, there are bikes to ride and it is all very nice and the children like it, but it is very much play rather than sort of subtle education, education without the children noticing. . . .   (Jane, Hallam)

Mothers in Southey Green, in contrast, are less used to talking about pre-school children's need for educational care. They express their children's needs for a pre-school education through a discussion of why their children need to go to nursery; however, most do not have specific views on what or how they want their children to learn. Some mothers, if pushed, talk about improving children's speech or teaching them to write their name, but in general mothers' knowledge and control of their children's pre-school education stop at the nursery gates. Only one mother articulated the desire for her children to go to nursery so that they could start a formal education. She was concerned that her own efforts were not only of limited success, but might be counter-productive:

I can't stimulate their minds, as much as I try with the ABC and numbers, basics, but I am not a teacher, you know I could be teaching them the wrong things. . . .
                                             (Alison, Southey Green)

Mothers' cultural and material resources: the importance of local social networks
These different attitudes towards pre-school education are reproduced in different ways in the two case-study areas. In Hallam mothers draw on a wider variety of information sources, including the advice of childcare manuals and health visitors, but formal and informal contact among mothers seems to be the most important way in which 'knowledge' about child development and pre-school education is disseminated. Many mothers, particularly those who work part time or are not in paid employment, consciously construct support networks for themselves by continuing links initially made at ante-natal classes or by using (and sometimes running) baby and toddler groups. These networks operate mainly during the week, with the weekend being defined as family time:

[At the] weekend again, everybody we know tends to sort of close ranks and it is a family time, so I don't see my girlfriends at weekends, because they are sort of remembering who their husbands are, 'yes, I seem to remember you were around at some point in my life!'   (Jane, Hallam)

This active and formal networking is particularly important for first-time mothers (and those new to the area): toddler groups for example, provide everything from advice on nappy rash, through information on childcare services, to emotional support:

That is another thing you see, you pick up from toddler groups and baby groups and that, you know that's what you should [do], other things available, or opportunities or things you could be doing with your kids, you know, it is not just a social thing you get loads of information and feedback from other people.

(Clare, Hallam)

Local mothers (and one father) also produce a free magazine called *Chatterbox*, which provides lists of services available locally, as well as information on possible days out, tips for the home and forthcoming events. It is distributed through the local women's hospital, doctors' surgeries, libraries, toddler and playgroups, etc. The magazine is particularly important in drawing new mothers (both those with new babies and those new to the area) into the local childcare system: several mothers reported having been given a copy of the magazine when they moved to the area; and the magazine itself lists the phone number of a mother in each district who is willing to talk to and welcome new mothers. The production of the magazine is dependent on voluntary work in terms of collating information, producing the magazine and distributing it to locations across the ward, and advertising space is sold to cover production costs. Like any voluntary effort the magazine has been through periods of insecurity, but at the time the fieldwork was undertaken it had been taken over by a new co-ordinator who had streamlined its production:

What I found [when I took over] was that, the previous approach had been everybody muck in and do everything, but then it meant that nobody felt important, nobody felt that it mattered and they just all felt overwhelmed because it was all too much to do really, so what I did was I looked at it like as a manager and I broke it down [. . .] and then sort of having like divided everybody up . . . I found that everybody felt fulfilled because they knew that, that was their little job and if they didn't do it, it didn't get done, you know, and it left everybody else in trouble. And that worked brilliantly and what I tried to do was build it up by areas as well. . . .   (Lydia, Chatterbox co-ordinator, Hallam)

Southey Green mothers' social networks take a radically different form. Many mothers here, like some in Hallam, have family living locally whom they see on a regular basis during the week. Mothers' own mothers are often an important part of these networks: Diana for example, sees her own mother roughly four times a week and her partner's mother once a week, but rarely talks to anyone else. A minority of mothers use toddler groups, and those who do can find them an important source of support. However, these groups are much less successful in this area, both because of perceived differences in respectability between areas within Southey Green, and because some mothers find the social tensions with the groups, which are often framed by judgements about appropriate femininities, off-putting:

Andrea: I don't think there's many people seem to take advantage of it [the toddler group] anyway.

Susan: Well what it is, people have thought that it's, you had to be in [the] clique, they thought it were a clique, but it weren't.

Andrea: It weren't. I mean there is and there isn't. There is them and there is us. Now us is: we'll have a laugh, we'll talk about anything, and we mean stuff like *anything*, and then you've got other side that are 'tutting', but nine times out of ten people that are at other side tutting are having a good giggle as well, aren't they? [. . .] there used to be one lady, no names mentioned, and she were right undecided on where she was, she didn't know whether she should have been with these people that, you know, have got fluffy slippers and go round with a duster, or with us that will sort of go round with Doc Martins and hoover and dust you know, and she really was unsure and she did it herself, she find it hard herself to fit in. . . .   (Joint interview, Southey Green)

Overall, women in this area tend to have fewer non-kin friends, as Cochran *et al.* (1993) showed was the case for working-class mothers in a variety of western countries, and some do suffer social isolation.

Educational childcare provision
In this context, knowledge about the importance of sending children to nursery and information about how to obtain access to a nursery place is passed on through family and institutional networks in Southey Green, as mothers draw on their own and other family members' experience of attending nursery school and on the advice of health visitors and sometimes social workers. Places are relatively freely available in nursery classes attached to local primary schools because Sheffield City Council targets its limited funds for nursery education on less privileged areas such as Southey Green. The pervasiveness of this advice, alongside the availability of places, results in the 'automatic' use of this provision and it is considered the 'good mother's' responsibility to put her child's name on a nursery list as soon as possible, and then to ensure that they take up the place when one becomes available.

The clear preferences that mothers in Hallam expressed for different learning environments and the wealth of information they hold at their fingertips might suggest that attitudes to provision directly governs their use. However, the characteristics of services such as their cost, timing, ages of children catered for and location play an important role. Moreover, underlying all of these service characteristics is the bigger question of availability. Sheffield City Council's policy of prioritising spending in less privileged areas means that no LEA nursery classes are available in Hallam. However, the cultural and material resources of many mothers in this area mean that they do know about a distant LEA nursery school, obtain places for their children (though fewer use this form of provision than those in Southey Green) and transport their children to and from nursery five days a week, for a morning or afternoon session. The use of the LEA nursery thus requires some knowledge of the system as well as the time and money involved in dropping and collecting children. Some mothers draw on their cultural resources, in this instance in the form of their social networks with other mothers, in order to minimise these costs. Josy, for example, used to share lifts with another mother when her first son went to nursery, and is organising this again for her second child:

[My eldest son] finishes [school] at 25 past 3, the nursery finishes at quarter past so I am in a better position to collect them every day, because her wee fella finishes school at quarter past and the nursery finishes at quarter past, and you can't be in two places at once. You know, so she will probably take them down every day and I will collect them every day. Obviously there will be days when that won't work, but, in fact, there is a girl who has moved in two doors down the road, she has a wee boy of that age and [my friend] is going to persuade her to go down and look at Broomhall and then maybe we will have three people to share the lifts with. It is just organisation really.   (Josy, Hallam)

The deficit of locally available LEA nursery provision has resulted in the presence of a vibrant playgroup movement in Hallam, which is largely absent in Southey Green. Playgroups are run by leaders (often mothers themselves) with involvement from the mothers whose children use these services. This involvement usually takes the forms of a mother-led committee which organises different aspects of playgroup life, including fund raising, and a requirement, more or less strictly enforced in different groups, for mothers to help in playgroup sessions on a rota basis. This involvement of mothers in the provision furthers one of the original aims of the playgroup movement, namely to teach mothers to mother more effectively. This ethos is still important for some groups:

There's always been a history of parental involvement, because it, we feel that, that side of it is very important really. I don't think they should come and dump their children without having much of an idea of what is going on because it can be quite educational for the parents, really, if they stay. If they have to come and stay they can see the sorts of things that children under 5 do. . . . If you have to come and help, you pick up all sorts of ideas. . . .
                                        (Margaret, Playgroup Leader, Hallam)

The experience of being involved in this way can increase the confidence of some mothers; however, the process should not be exaggerated as not all mothers get actively involved:

I wouldn't have the confidence to know, I mean I just, like I said, I have never been around children before and I don't feel very confident as a mother, never mind organising things for children to do. I wouldn't know where to start, I just don't have the imagination or the, so I really appreciate these women who do have the drive and the motivation and the ability to do all the things that I am using.   (Josy, Hallam)

The difficulties involved in using LEA nursery education and the availability of a market of working mothers and mothers not in paid work who are willing and able to pay for private nursery education has led to a rapid expansion in the number of private nursery places available within Hallam in recent years. These services are private businesses, not part of a corporate chain, and are usually run by local mothers who have identified a gap in their local market:

Why did you decide to set [the nursery] up?
Basically we both wanted to go back to work, I was an infant teacher and Janet
was a health visitor, and we both had three children under the age of seven,
and we looked at other nurseries, in fact we placed our children at other nurseries . . . and we both came away saying 'I'm sure we could do better than that'
[. . .] I do the older side of it and she does the under 2s, and people really like
the fact that it is professionals, not just anyone with money setting up and doing
it. That's where we've got the market differently from other people.

(Catherine, nursery owner, Hallam)

Configuring geographies and new femininities

Ideas about motherhood are central to many understandings of femininity, and
motherhood, as the quotation from Dally in the introduction to this chapter
suggests, is, like any other aspect of women's gender identities, socially constructed within specific times and places (see Chapter 1). The material presented
in the previous section illustrates how mothering practices, in particular in relation to questions about paid employment and the education of pre-school children, are shaped within the local childcare cultures of Hallam and Southey Green.
These contemporary social constructions of mothering sometimes directly
reproduce 'traditional' ideas and practices. Some mothers in both Hallam and
Southey Green for example, chose to replicate 'traditional' patterns of mothering by mothering full time in the home. Other mothers, however, forged new
femininities by combining motherhood and paid employment to a greater extent
than has been the case in the past. Here, mothers in Hallam, who had a better labour market position and were better able to pay for childcare, were more
able than those in Southey Green to combine the roles of mother and worker.
Differences between the middle-class mothers of Hallam and the working-class
mothers of Southey Green were also evident in ideas about the mothers' role
in pre-school education. In Southey Green mothers continued 'traditional' patterns of care and sent their children to LEA nursery provision; in Hallam growing numbers of women, not all of whom worked, used private and voluntary
as well as state-sector educational provision for their children. In this section
I want to draw out the importance of geographies in the construction and contestation of these 'traditional' and new femininities.

First, it is important to stress that these local childcare cultures mapped out
in the previous section are not simply local; they are porous spaces through
which national and international processes are both experienced and contested.
The changing position of Britain in the world economy and the feminisation
of the labour force, which we discussed in Chapter 2, are important here. This
can be seen in the ways some mothers, especially those in Hallam, are drawn
into the paid labour force and others are marginalised from it, particularly in
areas of higher unemployment such as Southey Green which have been hit hard
by economic restructuring. The two study areas are also shaped by national and
local social policies, with, for example, the lack of childcare for working parents reflecting national government's decisions to regard the care of pre-school

children as a private responsibility, and the different levels of nursery education in the two areas, in part reflecting the local authority's decision to prioritise spending in less privileged areas. Similarly, nationally hegemonic understandings of femininity which, as we argued in Chapter 1, have been dominated in Britain by the association of women with the home, have continuing influence in both Hallam and Southey Green.

Equally, while global, national and city-wide economics, politics and cultures have an important influence on these two areas, so too do the ways in which local mothers respond to them. This can perhaps most clearly be seen in the renegotiation of the local moral geographies of mothering. I have argued elsewhere that local childcare cultures have two (mutually constituting) components – the local moral geography of mothering and the local social organisation of non-parental childcare provision (see Holloway, 1998b, for more details). Local moral geographies of mothering are the locally constituted sets of ideas about 'good mothering', the highly localised discourses concerned with what is right and wrong in the raising of children. These moral geographies of mothering can be seen, for example, in the local understandings of a 'good mother's' role in ensuring her child receives a suitable pre-school education. In Hallam 'good mothers' must know a great deal about the available service types, evaluate these and ensure that their children receive the best form of care possible; in contrast, the 'good mother' in Southey Green is proactive in ensuring her child's name goes on a nursery list and takes up the place when it becomes available.

These moral geographies of mothering are more than just interesting examples of cultural diversity shaped in the context of different local circumstances. As the research demonstrates, the different moral geographies of mothering in Hallam and Southey Green are important in defining mothers as a social group and in influencing the meaning and experience of motherhood for individual women. To expand, the locally specific definition of 'good mothering' shapes the boundaries and rules of group membership. As Philo (1991: 16) has argued:

> it would appear that moral assumptions are crucially bound up with the 'social construction' of different human groupings – with deciding the character of these groups; with laying down the codes that groups live by, particularly in their dealings with others – and this means that spatial variations in everyday moralities will inevitably be closely entangled with spatial variations in the 'structure' and 'functioning' of human groupings.

Some women actively draw on this local moral geography as a resource to inform and validate their actions, for example comparing their mothering practices and childcare choices with those of family and friends. Other women who might not initially choose to mother in this way find themselves under pressure from other mothers and childcare professionals to conform (cf. Valentine, 1997a, 1997b). For example, a mother who does not register her child for nursery in Southey Green is likely to find herself under pressure to do so from family and professionals such as health visitors and social workers. Thus these moral codes influence individual women's experience of motherhood as they define themselves (and are defined) in relation to it.

However, it is important to remember that these local moral geographies of mothering are not forever fixed, but as socially constituted systems are open to contestation. Indeed, the case study clearly illustrates how women reworked ideas about 'good mothering' in the changing economic circumstances that make paid employment a necessary, and often valued, part of some mothers' daily lives. This reworking of 'traditional' femininities took a different form in the two case-study areas: in Hallam an emphasis on child-centred activities and part-time work came to be seen as 'good mothering', as this limited the time mothers spent at work and/or allowed them to 'make up' for their absence from the home; in Southey Green part-time work came to be seen as a 'good mothering' strategy because women were going out to work in order to improve their children's standard of living. These new understandings of 'good mothering' became important in validating some women's decision to return to work. Thus it is clear that the local moral geographies of mothering both influence women's actions and are also contested by the actions of women forging new femininities in changing social and economic circumstances. Local moral geographies are then both constituted and contested in place, and are, as part of the local childcare culture, an important element in the geographies of those places.

The different means through which these local moral geographies of mothering are reproduced highlights the important role particular spaces, in this case the neighbourhood, can play in the construction and contestation of femininities. In Hallam, neighbourhood-based social networks were a particularly important mechanism through which ideas about 'good mothering' are reproduced. As was outlined in the previous section, many mothers in this area actively constructed a set of highly localised social networks after having children, specifically to support themselves as mothers. These networks helped working mothers, for example, by providing information on the range of childcare choices available, and in some cases by providing recommendations for specific childminders whom mothers therefore felt they could trust. Equally, the networks were important to mothers not in paid work and many used baby and toddler groups, especially with their first children, to find information about childrearing as well as emotional and social support for themselves. Some women's community activism in this area – for example providing and running playgroups, acting as points of contact for new mothers, producing the *Chatterbox* childcare magazine – was important in reproducing ideas about women's roles as mothers and the neighbourhood as a family-friendly space. In Southey Green the concept of neighbourhood, in the sense of local community rather than physical, mappable space, appears at first sight to be less important. Differences in local interpretations of appropriate femininity – for example in whether single mothers are made to feel welcome at toddler groups, or how 'traditionally' feminine and demure women's behaviour should be at these groups – led to tensions that reduced the importance of such mothering groups for some mothers here. Nevertheless, many women still have very strong local social networks forged through familial rather than non-familial links, which meant ideas about 'good mothering' were freely circulated in the area. The importance of these networks between women and their families highlights the continuing importance of neighbourhood and community in middle- and working-class life, and in particular as a space through which domestic femininities are reproduced.

The home, and its relation with paid work spaces, is equally important as a site through which gender identities are made and remade. The home has traditionally received less geographical attention than paid work, though feminist geographers have for some time tried to elucidate the links between the two (Hanson and Pratt, 1988) and to retheorise our understanding of home:

> The notion of home is an exceedingly complicated one for feminists and it is one that feminists actively try to complicate: to unmask power relations within households, and to breakdown dualisms between home and work, production and reproduction, public and private. (Dowling and Pratt, 1993: 464)

The question of home is particularly important in discussing mothering practices in Britain because, as we suggested in Chapter 1, hegemonic understandings of femininity have worked upon the basis that a woman's place is in the home, and the dominant childrearing ideology of the post-war period has stressed that home care is best for young children. These ideologies shape our understandings of the home as a feminised and child-oriented space. Some women in both Hallam and Southey Green believe wholeheartedly in the importance of home and the essential role a mother can play within it, and therefore choose to mother full time in the home. Some others who go out to work, try to reproduce this home environment by using childminders and less often nannies to care for their children.

Indeed, Gregson and Lowe, in their study of nanny-employing households, argued that:

> the existing level of quality, accessible and widely affordable collective childcare provision in contemporary Britain permits only limited development of alternative non-home-based childcare spaces. There is, quite literally, only limited space for resisting the spatial prescriptions of the dominant ideology of childcare. Home space remains in many sense unchallenged as the space for young children in contemporary Britain. Small wonder that locating childcare in the parental home continues to matter so much to the middle classes. (Gregson and Lowe, 1995: 231)

However, this case study not only shows the enduring power of home in gender and childrearing ideologies, it also starts to show breakages and reinterpretations of the meaning of home. As was noted in the previous section, the availability of affordable collective childcare in Hallam, and the emphasis on the importance of educating pre-school aged children has led some middle-class mothers to challenge the assumption that 'good mothering' means mothering in the home and instead to place greater value on the stimulation collective-care environments can provide. In challenging the ideological link between 'good mothering' and mothering in the home, the meaning of home is also challenged. Ideas about spaces are not only important in the constitution of 'traditional' gender identities: our ideas about these spaces can be changed as women forge new femininities.

Conclusion

This chapter has drawn on case-study research in two areas of Sheffield, Britain, in order to examine some of the complexities involved in the reproduction of motherhood. Focusing on two issues in contemporary Britain where ideas about 'good mothering' are perhaps most clearly articulated, it shows how 'traditional' ideas about 'good mothering' are reworked in changing social and economic circumstances where more mothers are being drawn into the labour market, and where mothers as a group are increasingly being held responsible for the education of their pre-school age children. Through an analysis of the local childcare cultures in Hallam and Southey Green, and in particular a focus on the local moral geographies of mothering, the chapter demonstrates the importance of place-based cultures in shaping mothers and mothering, at the same time as recognising the ways in which some mothers actively contest these ideas in forging new femininities. This focus on local moral geographies also highlights the importance of the neighbourhood, and in particular local social networks, as spaces through which ideas about 'good mothering' are shaped and passed on. Moreover, in drawing out the importance of the home, both as a central site in many mothers' lives, and as an ideological construct of great force, the research illustrates how gendered discourses in part make spaces, and how our ideas about spaces can be changed as women negotiate new femininities. The research thus highlights the importance of spatiality in the reproduction of motherhood, as women negotiate 'traditional' femininities and forge new ones. The lessons this suggests for geographical thinking are considered further in Chapter 8.

CHAPTER SIX

Contested territories: women's neighbourhood activism and German reunification

FIONA M. SMITH

Introduction

This chapter explores how women negotiate their involvement in political action in the context of the forms of rapid change that have marked the 'transitions' from communism in eastern Europe and the post-Soviet states. The processes of political, economic, cultural and social change that constitute these 'transitions' have been discussed already in relation to issues of globalisation and neoliberalism (see Chapter 2). One of the key issues that this chapter seeks to question is the way in which dominant understandings of post-communist 'transition' and transformation regard them as undifferentiated, global-scale processes that 'impact' on passive spaces and people (Smith and Pickles, 1998). In the specific case of German reunification, this is reflected in the way the unequal power relations between 'East' and 'West' have become mapped onto the opposition between an active 'West', promoting change and providing the expertise, and a passive 'East', the recipient of change. Kathrin Hörschelmann (1997) argues that this move feminises the whole of the 'East' and contrasts it with a masculinised 'West'.

The implications of transition processes for women across the former communist states vary considerably, although there is agreement that women bear much of the brunt of restructuring (see Bridger *et al.*, 1996; Renne, 1997). While recognising the extent of change that women face in their lives, and in particular the disproportionate effects of economic and social restructuring on women, this chapter focuses on women's political action in local neighbourhoods in one city to illustrate that processes of large-scale change, such as postcommunist transition, not only affect local spaces but are also negotiated and contested in and through the contradictions of everyday sites and spaces, a point further developed in Chapter 8. The chapter explores some of the often complex and contradictory ways in which women rework their femininities through the opportunities and limitations of changing local geographies.

As Chapter 1 showed, women in eastern Germany are faced with the need to negotiate their femininities in a situation where the dominant femininities of 'West' Germany and 'East' Germany had diverged considerably over the

40 years of the existence of the two states in opposed ideological systems and where, as in reunification as a whole, western forms are coded in dominant discourses of change as the 'norm' which should be aspired to. In contrast, 'eastern' forms are coded as out-moded or even suspect because of their association with the communist past. The complexity of this situation is explored here in relation to women's activism in the spaces of everyday life. This allows the focus of attention to be moved from seeking to identify 'the' answer for how all women in eastern Germany negotiate reunification and avoids positing women solely as 'victims' of such change. Instead attention is given to the multiple, fractured and even contradictory constitution of the femininities of eastern German women.

The chapter is divided into two main sections. The first discusses how women negotiate their involvement in neighbourhood action. It outlines how women positioned this involvement as part of a general move to use the 'rights' and possibilities of the new democratic system, and indicates the problems encountered where these theoretical rights collided with the practical difficulties of implementing them in reality. It goes on to describe how women gained new skills and expertise through their local political activism, connecting their actions to the local population and to wider political processes in the city and beyond. The second main section of the chapter asks how women selectively rework their femininities to include local political action in the context both of the dominant genderings of home, work and family and local action in the GDR, and of the gender-specific transformation of these spaces in the processes of reunification. In doing so, they variously draw alliances with other women, reinforce the association of women with family and caring roles, or point to the distinctive femininities women in eastern Germany bring to the politics of reunification. Before that, however, the chapter begins by introducing the historical and contemporary context of eastern German women's political participation and then outlines the case study of neighbourhood activism in Leipzig.

Problematic relations between eastern German women and politics

'Women's politics [in the GDR] were by and large politics "for" rather than politics "by" women' (Weichert and Höpfner, 1990: 199). Achieving the equality of women and men was a formally declared element of GDR state policy. In fact, on many measures of 'equality', such as access to education and involvement in employment, these formal goals were achieved (Nickel, 1990, 1993a; Schmude, 1996). The vast majority of women in the GDR were involved in full-time employment and most combined this with having families. The state introduced a range of policy measures that enabled women to combine family and work, including widespread access to childcare. By 1989, over 95 per cent of 3–6-year-olds attended kindergarten, and around 80 per cent of children under 3 attended an infant daycare centre (Nickel, 1992: 44). Afterschool care also provided for around 80 per cent of children in the lower years of school. The GDR's explicitly pro-natalist 'mommy politics' (Ferree,

1993) included free access to family planning, abortion on demand, substantial state benefits, credits on the birth of children and a guarantee that a woman's work-place would still be open for her after up to one year of maternity leave. These were designed to enable women to become full members of this 'workers' and peasants' state' by participating full time in the labour force while also producing and rearing the next generation of workers.

Many women appreciated the ability to combine family and career, seeing paid work *and* having a family as 'normal' elements of women's identities. However, they did so by performing a 'tightrope act [balancing] between motherhood and the occupational performance pressures of a planned economy' (Nickel, 1993a: 234), particularly since women continued to bear overwhelming responsibility for family and home. Striking the balance was a key part of women's lives (Böckmann-Schewe *et al.*, 1995). At the same time as trying to balance home and work, women's participation in the labour force was also highly gendered, often concentrated in stereotypically female sectors and particular occupations within other sectors (Nickel, 1992, 1993b). This vertical and horizontal labour market segregation resulted in women earning, on average, lower wages than men. Despite this, women's contributions to household incomes and to the national economy were substantial (Schmude, 1996). 'Equality' for women in the GDR was therefore achieved through an expansion of femininity to incorporate the spaces of home and work, rather than by renegotiation of femininities and masculinities. Furthermore, Christine Lemke (1991) describes the position of women in the GDR as one of 'equality without rights'. There were many fewer women in the structures of power within the country, and their proportion in relation to men declined as the level of power increased.

Not surprisingly, women's reactions towards this situation are somewhat ambivalent. One the one hand many writers suggest women recognised their abilities in coping with this 'double burden' while also rejecting the difficulties the situation caused them, a finding shared across many former communist states (Corrin, 1992; Funk and Mueller, 1993). Others suggest that, alongside their assumptions that paid work was a 'normal' part of their femininities, many GDR women retained rather 'traditional' views on the relation of women to family and home (Nickel, 1990; Lemke, 1991). This 'traditionalism' was both portrayed and reinforced in the images of femininity presented in the mass periodicals for women (Dölling, 1989) and in the femininities promoted by the official women's organisation, the Demokratischer Frauenbund Deutschlands (DFD) (see writings collected in Kahlau, 1990). This leads Weichert and Höpfner (1990) to argue 'equality' in the GDR remained based on the maintenance of social differences between women and men and the reinforcement of traditional roles and attitudes.

It was out of a critique of this 'state-patriarchal' system that the GDR feminist movement emerged (Weichert and Höpfner, 1990; Hampele, 1993; Penrose, 1993). Although some groups had previously worked in loose alliances, it was only in 1989 that formal oppositional women's organisations were founded. This, along with the considerable involvement of women in the other new parties and movements founded at this time, meant that women were very much involved in the politics of the 'revolution' period. However, while some of these organisations explicitly promoted women's or even feminist agendas, many others advocated the general goal of establishing a more democratic political system with no explicit reference to the position of women.

The subsequent 'transition' for eastern Germany has been defined by the process of reunification, a process that has introduced the pre-existing western German system of law, government, economy and social organisation, opening up eastern Germany's economy to global competition and privatisation. This has resulted in widespread restructuring of the labour market, with steep rises in unemployment. To soften the effects of this change, the federal government introduced a series of measures such as extensive training programmes, forms of workfare (*Arbeitsbeschaffungsmassnahmen*, or ABM) and transitional payments for early retirement. In virtually all these measures women make up the majority of those affected, and women also constitute around two-thirds of those registered unemployed but not involved in these measures.

As well as businesses, housing stock and real estate has also been privatised, principally through the restitution of property to its previous owners and sale to new owners (Blacksell *et al.*, 1996; Cassell, 1996; Smith, 1996c; Reimann, 1997). Market rents have been introduced and this, along with the restitution of state property to private owners, has particularly affected the housing circumstances for tenants who form the largest sector of the housing market. All of these processes are socially and spatially highly varied in their effects. Clearly, women's lives have been affected not only by the new possibilities of consumption, travel and democratic rights but also by employment change and by considerable change in housing issues. In addition, the provision of social and childcare services is changing considerably as industry sheds its previous responsibilities to provide them and the public sector faces a shortfall in funding. One of the few areas in which pre-existing western forms have not been introduced is in abortion law. Previously free on demand in the GDR, but illegal except on medical grounds in the Federal Republic, the debate around 'paragraph 218' produced one of the few areas of compromise between the two systems, but it nevertheless reduced eastern German women's access to abortion (Maleck-Lewy, 1995).

A series of authors have commented on the decline of women's participation in political action across eastern European countries since the activism of the various 'revolutions' (Einhorn, 1993; Nève, 1997; Watson, 1997). Reasons given include the legacy of the patriarchal structuring of political activity in the past, the complex ways in which women have reacted against what some saw as an enforced politicisation of life in the past (Tóth, 1993) and the fact that very few women were the main leaders of the reform movements (Penrose, 1993). These factors intersect with the highly uneven impacts of the economic problems of 'transition', particularly in terms of women's unemployment and difficulties in accessing social services such as childcare. However, some argue that while it is important to discuss these issues, too many studies fail to question the very definition of what actions constitute 'politics' and the scale at which political action can be examined. Regulska (1998) argues that this arises partly from the expectation among certain views of the processes of transition that significant change happens at the scale of the state. This, she observes, not only ignores crucial work in implementing and challenging the multiple processes of transition at the local scale but also excludes attention of the very areas in which many women are politically active. Such considerations led Bütow (1994: 268), reporting on a study of women in Saxony in eastern Germany, to conclude that 'one can only talk of the political non-participation of women if one bases the argument on current definitions of politics.'

One final point complicates any assumption that women's involvement in local activism can be interpreted in a straightforward way. The form of oppositional politics in the GDR (although limited in its extent) and in other communist states revolved largely around a notion of the possibility of opposing the state through the creation of a non-state, non-family 'civil society' (Benhabib, 1992; Fraser, 1992). In practice, in the GDR the state-security service's actions ensured that much of this work had to operate in close family and friendship circles. Peggy Watson (1997) argues that the very ideal of civil society as something that might escape from the home to become 'truly' political reinforced the dominant gendering of the home as a feminine sphere removed from politics and positioned all action outside the home as 'political' and therefore masculine. Not all scholars agree with this assessment, but it usefully points to a complex gendering of home, local space and political action in the GDR. This suggests that rather than assuming particular spheres are defined in specific ways, one should be open to the complexities of their constitution. The ways in which women in one eastern German city now negotiate the problematic gendering of different spaces in relation to political action in the processes of reunification form a key focus of this chapter.

The case study: neighbourhood activism in Leipzig

The material for this chapter comes from a larger study of political activism and urban change in eastern Germany during the early years of reunification (Smith, 1996b). It investigated the politics and developments surrounding change in five neighbourhoods in the city of Leipzig: Gohlis, Neustädter Markt, Plagwitz, Stötteritz and Waldstrassenviertel. All five neighbourhoods experienced, to a greater or lesser extent, the problems of housing decay, environmental degradation – particularly through the presence of heavy industry – and poor physical and social infrastructure that characterised much of urban eastern Germany. All lie within the nineteenth and early twentieth century parts of the city and largely exclude the massive housing estates constructed since the 1960s. After reunification and considerable local lobbying in some cases, at least parts of all five neighbourhoods were included in the list of 21 urban renewal areas designated by the city for extra support and attention. At the same time all were affected by processes of restitution and privatisation, employment change, reorganisation of social facilities and selective reinvestment (Wiest, 1997).

Leipzig's new urban politics were marked not only by the long and slow process of implementing a functioning western-style administration (Osterland, 1994). They were also strongly influenced by the continuation and further development of the civic politics of the 'revolution' period and the demand for more inclusive forms of urban policy development which arose from newly public debates in the early period of 'transition' (see Smith, 1999a). Out of this there emerged a series of civic action groups which concentrated on issues of urban development. Some were thematically based or worked on specific projects. Others focused on representing the views of neighbourhood residents and influencing local development. One such group was active in each of the five case-study neighbourhoods. Their membership included men and women from the local area, with women constituting between 35 and just over 50 per cent of the membership. It is women's involvement in these neighbourhood action groups (rather than in the specific 'women's' projects in the city) that is considered here.

The neighbourhood action groups sought to ensure residents' voices could be heard in the processes that were transforming the urban environment. In order to do so, they utilised and created public spaces for action in their local areas in ways previously not available to them. Public meetings on aspects of neighbourhood planning, housing law and social issues were held in schools and church halls. Public festivals were organised in neighbourhood parks and squares. They were used to develop the sense of local community and to publicise more widely neighbourhood planning strategies (see Figure 6.1). In order to facilitate their work, four out of the five groups also had an office where they could co-ordinate their actions and residents could seek advice and information or make suggestions about local development.

Groups saw themselves as organisations through which the interests and needs of the local population, of which they themselves were part, could be articulated and highlighted to city politicians, planners, developers and others whose actions were significant in changing local conditions. This took the form of specific campaigns on problems such as traffic crossings as well as longer-term involvement in developing resident participation in local planning. In addition, in all five neighbourhoods the action groups also set out to use such activities as local festivals, exhibitions of photographs or work with local children to increase the awareness of the local area among the population and to increase their sense of belonging. Finally, most groups also addressed local social problems with voluntary work from their members and paid input from ABM-programme personnel in such areas as support for the elderly or campaigning for children's and young people's facilities to stay open. The ways women make sense of their involvement in these groups are examined here by drawing on group discussions and individual interviews conducted with them between 1991 and 1993.

Figure 6.1   Creating public space through a local street festival, Gohlis, 1992.
© Fiona M. Smith

Learning the ropes: women's changing political subjectivities in neighbourhood action

This section explores, firstly, how women defined their involvement in neighbourhood action in relation to new political rights and possibilities of democratic involvement at the same time as anchoring them in the practicalities of the changes affecting their neighbourhoods. At times they also criticised the difficulty of realising these rights in the complex practice of urban renewal. The difficulties also led some women to identify more with being 'east' Germans and others to demand that the form of westernisation that was implemented should take account of the specifics of local circumstances. The section goes on to illustrate how women described their involvement as requiring them to develop and 'learn' the political skills necessary to make their local work effective.

(Re)locating and (re)defining political subjectivities

In the face of rapid and complex changes in the urban environment, most women argued it was important for local action groups to exist so that local people's concerns could be heard. They were keen not to allow either the city council and its administration or economic forces to change their neighbourhood without reference to their needs and wishes. For example, Frau Hönsch argued local activism would ensure that:

> a certain closeness to the citizens comes about, so that not only the adminis-
> tration gets to decide, but that in principle we can and may have an input too.
>
> (Frau Hönsch, Gohlis)

This, it was argued, required people to rethink the possibility of their involvement and to recognise the rights that they could use. In the context of the threats
of displacement of tenants from their homes after the introduction of new rental
conditions, one woman argued these rights would help people to stay in their
local areas. Another suggested this new form of politics could also be pro-active,
allowing local residents to contribute their policy ideas to the city, while a third
stated that involvement was preferable to being passive and letting things
happen:

> We want to give people courage, you know? To give people encouragement and
> to say, you have certain rights, don't let yourself be used and just quit the field.
>
> (Frau Dörmann, Waldstrassenviertel)

> Well, I think it is still really important that the people understand that they should
> and can get involved in the first place. Not just accept some change or another
> but that from the outset they try to make contact with the City, offer solutions.
>
> (Frau Tucholsky, Waldstrassenviertel)

> I wanted to be involved, because I think you cannot just simply wait and then
> let yourself be put out of your flat at some point. Rather you have to do some-
> thing about it, even if it is just limited means that we can use. [. . .] Just doing
> *nothing* and waiting – that isn't my nature and so I am definitely in favour of
> trying, especially in our citizen initiative, trying to change things.
>
> (Frau Berghahn, Waldstrassenviertel, original emphasis)

Many women therefore saw their involvement as a way of exercising the new
political rights that the change to a western democratic system had brought.
This allowed local concerns to be articulated in a manner that had been highly
problematic in the centralised system of the GDR:

> Well, I thought the agenda was really great – that the people who were really
> affected could get involved in their neighbourhood. That is precisely what there
> never really was properly before.   (Frau Schulz, Neustädter Markt)

In fact, as the following quotation indicates, some women had consciously
avoided involvement in the past and contrasted that situation with their willingness to take advantage of the new conditions which prevailed:

> I was never active before because I knew that I would just run up against brick
> walls and so I saved myself the trouble, you know? I . . . knew that I wasn't
> particularly capable of pushing things through like that. Only now after the
> 'change-over' [*Wende*] where I think that we should *use* the chances of a
> *democracy* – that's when I gathered up the courage to get involved and to fight,
> you know?   (Frau Dörmann, Waldstrassenviertel, original emphasis)

The problematic nature of the decision to be politically active was, for other
women, not confined to the GDR period. They described their decisions to

become members and active participants in local action as conscious rejections of being positioned as 'objects' upon which change would act. For them local action was a way of shaping their own, and their neighbourhood's future:

> Well, so, the two of us [looking at another woman], well we came here just to register, really just to sign up for the group and to be represented. And when we did, we were asked 'don't you want to get a bit more involved?' So then we thought about it and said to ourselves 'yes, at some point you have to do something, things won't happen by themselves.'   (Frau Herrmann, Neustädter Markt)

> I think there are two possible ways for us to go – the one is 'everything', and the other is 'nothing'. [laughs] [. . .] I think the whole thing is a bit schizophrenic. The people have the need – 'we must really all hold together, do something all together' – but on the other hand it is totally difficult to *actually* do something. Somehow under the surface, I think, there is always the feeling that it is no use. [. . .] That's true for me too.   (Frau Rolf, Plagwitz, original emphasis)

By locating politically active subjectivities in local areas, women not only defined this type of action as distinct from the hierarchical and centralised politics of the GDR which had excluded independent initiative and particularly local concerns from policy formation. They also clearly rejected being passive in the face of the restructuring associated with reunification. In defining local action as political, however, they nevertheless clearly differentiated it from party political structures and party dogma. Instead they proposed an expanded notion of political action which stood outside or above party political loyalties:

> Everyone thought – a citizen association – that's not something political, is it? Anyone can go there, it isn't a certain orientation, not PDS [reformed communist party] or SPD [Social Democrat Party], or even Bündnis 90 [civic movement].
> (Frau Schwarz, Gohlis)

> *Frau Dörmann*: We see ourselves as . . . well, outside, or rather – what's it called? Not political, what do you call it?
> *Frau Schulz*: A small stone in a big mill?
> *Frau Dörmann*: I can't get the right word – above party politics! [. . .] We don't want to be in favour of one party or the other. [. . .] It is about our overall goals here and not about some party membership, you know.
> (Group discussion, Waldstrassenviertel)

This constituted both a reaction against the dominance of party ideology in the GDR and an attempt to find a space for politics in the new system that did not have to conform to the new structures of party politics. Some of the groups had, in fact, been established in the interim period between the collapse of communist rule in late 1989 and reunification with western Germany in October 1990. However, members of all the groups connected their actions with some elements of the politics of the civic movement that developed in 1989/90. (See Smith, 1999a, for an account of the overall development of Leipzig's urban politics in relation to that period.)

At the same time, around 30 per cent of the women members of the five groups who responded to a questionnaire survey in 1992 indicated they were also members of political parties, a level considerably above that for the population as a whole, and several group members were also local elected politicians. Some of the women were therefore also politically active in other forms. Nevertheless, there was a clear, and in most groups constitutional, understanding that action in neighbourhood groups should depend only on local residence or a concern with the local area. The refusal of any formal ideological basis for such action marked these groups out as a space for action that was seen as free from the limitations of party politics. Their legitimacy was based on the claims of local residents to exercise their new rights and, as two women who were less involved but still members argued, neighbourhood groups were regarded by the local population as 'representative' because their members were fellow residents who shared the concerns of the local population:

> I think most people who live here do not participate but they do follow what is going on and they know, it comforts them, that there is someone there who represents *our* interests.   (Frau Vogel, Neustädter Markt, original emphasis)

> There's this committee and a core group of people. And I really do think that they do it so well that it is representative, that the citizens recognise themselves in it. It isn't some kind of distant troupe which is doing something but people who all live right there in it too.   (Frau Tucholsky, Waldstrassenviertel)

Women's involvement in activism therefore drew together both more universal notions of the democratic rights of citizens and specific notions of the value and validity of local concerns in the face of the widespread processes of change that faced these neighbourhoods. There were, however, situations where the women were highly critical of the ways in which the circumstances of change came together to prevent both of these ideals from being realised. Specifically, several women argued that the length of time that the planning process took, combined with the random effects of the property restitution process in blocking proposed improvements on particular buildings (see Smith, 1996c; Reimann, 1997), left them feeling that in some respects 'the situation is no different from what it was before' (Frau Herrmann, Neustädter Markt). They were again faced with the promises, delays and disappointments to which they had become resigned in the GDR:

> You know, everything was always delayed in the plan economy [laughter]. But now [. . .] nothing has happened. And the disappointment is very great.
> (Frau Friedrichs, Neustädter Markt)

> For example with my building [. . .] renovation was halted after only the roof had been done because this restitution claim arrived. No one will now touch it with a barge pole.   (Frau Gerster, Neustädter Markt)

> Before [in the GDR] a lot was promised. [. . .] I know that *my* building, for example, was supposed to have already been renovated in 1980. It is still waiting today. [. . .] And that, that makes the people so depressed because they have only *ever* heard promises and the promises are continuing.
> (Frau Schulz, Neustädter Markt, original emphasis)

Criticism was also directed at what was seen as the inability or unwillingness of the city authorities to listen to local opinion once plans moved from the general principles of how local development should proceed to specific issues (see Smith, 1999a, 1999c, for more general discussions of these problems). While in 'theory' the right to involvement existed, several women suggested there were limits in the extent to which they could realise these rights, something which Frau Johansen, for example, suggested was reminiscent again of the exclusion of people from decision-making in the past:

> It is just a different front from what it was before. You have the right to parti-
> cipate [*Mitspracherecht*] but in the end that only works in, at particular points.
> (Frau Johansen, Neustädter Markt)

> On the one hand there is the theory and the legal basis that certain steps
> have to be followed and that the citizens have to be asked and that their right
> to involvement is to be guaranteed. Rights that are positive. But practically it
> isn't enforceable.   (Frau Herrmann, Neustädter Markt)

Women therefore criticised the lack of change and in particular any circum-stances that left them feeling that the western form of development did not produce an improvement on the previous system. Their articulation of local concerns often drew attention to the shared problems facing people in eastern Germany. They also criticised the form of development taking place, which apparently prioritised the rights of property owners over local residents, as one that failed to pay attention to the particular conditions facing eastern Germany:

> You cannot simply apply the structures that apply in western Germany to what
> is happening here.   (Frau Töpfer, Neustädter Markt)

At times, however, they also drew parallels between their civic activism and actions which they knew of in western Germany to draw attention to the fact that western forms of development were also highly contested in themselves:

> There are in the Federal Republic, or Old Federal States [in western Germany]
> projects where they have tried out forms of renewal where the citizens are closely
> involved too.   (Frau Töpfer, Neustädter Markt)

Developing skills, making links

Women's understandings of their new political action were confined neither to abstract conceptualisations of democratic rights, nor to lists of the failings of the transition to democracy. The project of realising these rights and mak-ing change happen operated instead through a range of very practical experi-ences as they negotiated for themselves the parameters of local action. The new and complex situations of which none of the participants had any prior experi-ence meant women spent much of their time negotiating what types of action they could and should undertake. Women commonly described the period of starting up neighbourhood action groups as one involving considerable effort in learning about how local activism could operate. For example, in one area of run-down nineteenth century working-class housing women campaigned to make the city planning department and others aware of local conditions. The pride they felt about this work arose from the fact that it was based on their own initiative. Doing this political work required that they also, quite literally, find their way around and get to know their local area:

> We decided to make clear to the people responsible just how bad the living conditions were for people here. It was really our first fight at that time. We walked around the area here and looked at everything and wrote to the mayor and tried to make it clear to him [. . .] that there were extremely poor conditions, where the kitchens of ground floor flats were sometimes subsiding into the cellars and were still inhabited. And we managed – it was of course only a small step – to get three families re-housed. [. . .] We were so proud of that, because we were in a very poor position because no one, no one was concerned about anyone at all [the city departments were all in the process of complete change-over too]. [. . .] We think that we *did* achieve some things, above all the recognition by the city departments. You'd really have to say that we achieved a certain standing at that time.   (Frau Herrmann, Neustädter Markt, original emphasis)

Women members were clear that the lack of experience and organisation in the city administration presented both problems for getting things done in their areas but also possibilities for them to influence policy formation before the city's administration and its policy formation structures had become too fixed. However, several women suggested that a similar learning process was necessary in their own activism. As the following quotations show, many of the women felt they had not really known how things would work out and had been 'innocent' and 'curious' when they joined:

> Well, I myself came to be involved in the same way that a virgin gets a child. I was simply *curious* and followed up on the notice in our building which [. . .] awakened my interest because I just said to myself, I have lived in this quarter for a long time already and I'm interested to see that something comes of this quarter, that it isn't destroyed.
> (Frau Dörmann, Waldstrassenviertel, original emphasis)

> I never knew how it would all develop, that it would all get so serious. [. . .] It was just simply a totally spontaneous decision coming from the whole situation, from this radical change [*Umbruch*], to really do something . . . that there would be great stones blocking our way on all sorts of things, and the administration which we have to do now – we had no idea.
> (Frau Schulz, Neustädter Markt)

This early period represented for some a particularly exciting time when local action was not yet caught up in the regulations applying to employing ABM staff, for example, or in the formal requirements for citizen participation in the new systems of planning for urban renewal which reunification brought. They also suggested that although their work had not conformed to the rigorous standards of data collection which were later demanded of inputs to the planning process, they had nevertheless produced useful information:

Frau Friedrichs: We did that first social survey out of ignorance [laughs]. Well in 1990 – really!
Frau Ziegler: Up stairs and down stairs.
Others: Oh yes!
Frau Friedrichs: Not like it is supposed to be, like, systematically, but just asking at the doors – how many rooms they have, and how the flat is, and what they think and whether they want to stay living in the area, how many men and women and children under 18 [. . .] and then we took all these opinions and evaluated it ourselves.   (Group discussion, Neustädter Markt)

Frau Friedrichs in fact suggested that the engagement of ABM staff to carry out some of the group's work had subsequently produced a greater distance between group members and the work in the local area. In contrast, she characterised the earlier work as 'real' grass-roots activism:

I'm *not* undermining the last social survey that was done [by a trained social scientist on an ABM contract] – but we aren't really all involved in that. [In 1990] it was real grass-roots work, just to, just even so we were able to argue with the authorities, with the departments in some kind of way.
                    (Frau Friedrichs, Neustädter Markt)

Over the subsequent two or three years, many women felt they had learned about the complexities and possibilities of the processes of reunification across a range of issues, including housing law and tenancy rights, the environment, planning procedures and social and cultural issues of local significance. Since they recognised that their own need to learn reflected a more general lack of information and experience of the new situation, they argued one of the most important actions that groups undertook was to provide forums where residents could access this information. Most were involved in forms of 'citizen advice' or in organising local public meetings with specialists, particularly in the area of housing rights:

Frau Dörmann: At the moment the most important thing, I think, is that we enlighten the people a bit, how things work, and how things shouldn't be allowed to happen.
Frau Lang: Yes, it is a very important point, because *we* don't know our way around the new legal situation at all. At a public meeting they can find out how to use their rights and that you can't just put tenants out of their homes.
                    (Group discussion, Waldstrassenviertel)

Knowledge of issues was, however, not sufficient. Groups also had to establish useful forms of political action and influence. Various women argued that the groups had to establish the right types of connections both within and beyond their local area. Gaining recognition among city officials required that the groups work to create good working links to them and to demonstrate their ability to consider the advantages and disadvantages of various policy options. This demanded consideration of the issues rather than quick reactions against proposed changes, as the leader of the group in Neustädter Markt suggested in relation to a local traffic planning issue:

We held discussions, we went there and presented our viewpoints. [. . .] And those are things you can only argue about when you have already concerned yourself with them for a while and thought about the pros and cons. It doesn't work when someone who lives in a street and is just annoyed by the plan goes along and tries to veto it. [. . .] And that was . . . not exactly a fight, but it took a time until the citizen initiative *itself* was recognised in the city departments.

(Frau Gerster, Neustädter Markt, original emphasis)

In developing these skills to deal with the authorities, the women saw themselves and the groups as having gained recognition and a level of expertise in political negotiation. They also suggested that the groups had developed a form of activism that did not simply seek to displace problems to other locations but considered different solutions and their effects on the neighbourhood and the city as a whole. However, they stressed not only that they had the right to influence planning but that the planners should also utilise the value of residents' local knowledges, particularly where many of the 'professionals' had come from western Germany and had little knowledge of local conditions. In a plan for traffic calming in Waldstrassenviertel, for example, a committee member in the local group, Frau Dörmann, suggested this local knowledge considerably improved the plans:

The plans were presented to us and we sat together round the table with the Traffic Department and discussed and came to the conclusion that not everything which worked on the plan was possible in practice. And in such situations, as residents of the quarter, who often know more about it than some planner from, say, Osnabrück, or [another western German city] who isn't so knowledgeable about the situation here.

(Frau Dörmann, Waldstrassenviertel)

Local activism therefore brought recognition of the value and validity of local concerns, drawing the neighbourhood and the actions of its residents into the processes of policy-making. Various women argued it was important to form wider alliances with the press in the city, which was largely positively disposed towards local action groups, or with the array of like-minded groups across the city:

When things are going against our plans, or where there are problems which we can't solve, where we can't find a common line with the departments then we *have* to make it public. So we can prepare some things or invite the press along to particular events and get things publicised in that way.

(Frau Gerster, Neustädter Markt, original emphasis)

When everyone gets together [. . .] then 64 associations and initiatives can really find *strength* together. [. . .] And when such massive pressure comes from the mass of people, that can achieve something, you know? Or with the exhibition on 'Citizens shape their city' [held in 1992 by all 64 groups] – I think that really made some people think and take notice: 'Aha! It's amazing all the things that are possible.'

(Frau Dörmann, Waldstrassenviertel, original emphasis)

Overall this and the previous sections have shown that women worked hard to develop particular forms of local action, exploring new possibilities that had not been available to them before and, through a complex variety of strategies and tactics, articulated the concerns of their neighbourhoods in the context of the restructuring facing them during reunification. In general, women positioned this political activity in relation to an ideal of civic politics and the newly available democratic rights. Generally they did not differentiate a specifically female agenda within this form of politics. Divisions within the identity of 'citizen' were more common along the divisions between 'East' and 'West' than between women and men within their neighbourhoods. However, this does not imply that women's femininities were not changed by these processes. The work of (re)positioning themselves, as active and increasingly knowledgeable citizens, and their local and everyday spaces, as valid scales for political action in relation to both the city and broader processes of reunification, disrupted any attempt to exclude spaces typically gendered as feminine and the everyday concerns of these women from the realms of political concern. In the process, both women's identities and their geographies were reworked to incorporate new forms of political involvement.

Reworking local political femininities and their geographies

The involvement of women in these forms of local activism is not an aspect of these women's identities which exists in isolation from other elements of their femininities. This section explores in particular how women were engaged in drawing on and reworking aspects of their femininities and the significance of the gendered spaces of the neighbourhood, work and home. This reworking occurs both in relation to the past in the GDR and in relation to the processes of reunification.

In the past in the GDR, the neighbourhood was somewhat ambiguously gendered. It was treated in official policies as a realm where reproduction predominated, outside the core site for political and economic organisation in the work-place (Höhmann and Seidenstreicher, 1980). Simultaneously the state's fine-grained structures of social and political organisation acted as a social technology extending state and Party influence to the most local of levels. The ruling communist party, the SED, organised primarily in the work-place, with only those members excluded from paid work through retirement or the like organised in neighbourhood cells. However, it formed the *Nationale Front*, together with the GDR's other parties and the mass official 'societal organisations', such as the GDR youth movement (FDJ), and the Democratic Women's Federation (DFD). This union of 'societal interests' organised neighbourhood committees (Wohnbezirksausschüsse, WBAs) which covered the whole urban area (Lapp, 1988) and their task was to encourage the local population to be involved in a variety of social campaigns.

In practice, levels of activity by the WBAs varied considerably and the majority were most active at the times of the GDR's elections. More significant in many areas were the local groups of the various social organisations, such as the DFD or the FDJ, or the *Volkssolidarität*, a welfare organisation particularly for elderly people, as well as, for example, parents' committees at schools and kindergartens. All were groupings in which women were heavily involved. Together with the lower significance attached to the neighbourhood, the predominance of women in these organisations reinforced the 'feminisation' of the local sphere and women's identification in local action with a variety of 'caring' roles. However, neighbourhoods were simultaneously sites for employment for many women. Fifty-seven per cent of the women surveyed from the groups said their current or most recent job was located in their neighbourhood. Given the close proximity of housing and industry in the old areas of GDR cities, this level is perhaps not surprising. However, Lemke (1991) also points out that women in the GDR often chose their place of work precisely in order to be able to combine their family and work activities more easily. Nevertheless, local areas did represent a space in which many women were socially involved.

While we have seen above that for many women the neighbourhood action groups founded after 1989 represented new forms of activity, there were also a significant number of women (25 per cent in the questionnaire survey) who had 'always been involved in working for the neighbourhood'. These women spoke of how these social and cultural activities had provided them with local contacts, particularly with other women in the neighbourhood. Some felt the loss of these structures of social organisation had contributed to a decline in levels of local collective involvement:

> I used to be in the DFD, you know? And through that I knew lots of women . . . we set up lots, eh, lots of things with the women and the children and so on . . . before the *Wende* still. [. . .] I mean, there was something almost every month. But now it has become markedly quieter. [. . .] In my opinion there were considerably more organised events before where people could volunteer to take part.   (Frau Hillmann, Waldstrassenviertel)

These earlier forms of neighbourhood involvement had been based around the willingness of women, particularly 'mothers', to be involved in a range of local activities:

> *And who was responsible for organising these local events?*
> *Frau Hillmann*: Mostly the women themselves . . . and over the years there were also the ones who had children – they were in the school parents' board.
> *Frau Berghahn*: There were holidays organised too.
> *Frau Hillmann*: And when there was something to be collected for, for all sorts of things, like for Africa or whatever. I mean, there were, eh, you could always ask the women who were involved in the school. [. . .] How often did I help with school parties, children's carnival and all sorts? [. . .] When they were having a party, the teacher would ask if the mums who had a garden could donate jam or jellies, or whatever.   (Group discussion, Waldstrassenviertel)

Women being active individually and collectively was, therefore, not an unusual phenomenon here. However, these discussions also confirm the extent to

which women rather than men were those responsible for children and local
social care in the GDR. In particular the women largely uncritically equate parenting with mothering, a point to which we return later. In the discussion around
what might have changed since then, this gendering of certain forms of action
remains unquestioned. What is questioned is the extent to which the previous
forms of action were independent of state structures. Frau Dörmann, for
example, argues that a new form of action is now required, distinct from the
ways in which social actions were organised in the past, and suggests that this
necessitates to some extent people rethinking their own relation to such action:

> Then, before, a lot was just ordered from above. Perhaps that was sometimes
> not all bad. Sometimes it missed its target, although a lot did come out of it.
> [. . .] Many people have been led by the hand their whole lives – 'now do this
> and now do that' and they joined in. [. . .] Now you need more of your own
> initiative. Lots of people still have to learn that. [. . .] But that is something where
> we, where we have to think about our *own* actions.
>
> (Frau Dörmann, Waldstrassenviertel)

Bearing this in mind, the question arises as to how women involved in action
groups after 1989 drew on and reworked the gendered spatialities of the
home and the neighbourhood as they incorporated the new forms of political
activism in their subjectivities. For many women, personal circumstances,
family issues and attachment to their 'home' neighbourhood provided motivation for becoming involved in the new neighbourhood action groups:

> A friend took me along [. . .] and she said to me, now we have to find out
> what will happen to our flats and everything around about them now that the
> market is coming.   (Frau Hillmann, Waldstrassenviertel)

> I've lived in the quarter here for 37 years and, well, after the *Wende* there were
> rumours that [. . .] we might have to move away from the Waldstrassenviertel.
> And that sat in my heart like a nightmare [sigh], I have to say.
>
> (Frau Lang, Waldstrassenviertel)

> My aim was really to be part of the formation of the neighbourhood in which I
> have lived for 20 years and because I have children and grandchildren [. . .]
> who live here too.   (Frau Hönsch, Gohlis)

> It has been my home since I moved here after we got married [. . .] and if this
> could once again be a nice corner in Gohlis, it would be really nice.
>
> (Frau Schwarz, Gohlis)

As we have seen earlier in this chapter, however, women's concerns in local
action were both personal and more general. By involving themselves in collective forms of action, the women extended their concerns with the personal
and the home to more general concerns with the neighbourhoods in which they
lived and (often) worked or had worked. This sense of concern for a wider
spatial and social community was echoed in many women's identification of
the need to help sections of the local population for whom the period of change

raised particular difficulties. Typically the groups most affected were seen as the elderly, and other less well-off households, and young people:

> There are many, many older people here, pension-aged, many single people. [. . .] We said, at some point you have to do something for these people.
>
> (Frau Herrmann, Neustädter Markt)

> The young people will be left high and dry if the leisure centre in Leibnitz Street closes. And somehow we have to find another solution. [. . .] Where are they supposed to go?   (Frau Dörmann, Waldstrassenviertel)

In response to these issues, many groups specifically set out to develop relevant projects and campaigns, often staffed through ABM posts (the majority of whom were women) but also involving volunteers. In one sense, then, these concerns represented a continuity of local action in the GDR where women's voluntary actions were particularly associated with the social and welfare sectors. In the situation of increased non-employment, particularly among women in the 40–60 age group, women could maintain a socially significant role through local action:

> *Frau Berghahn*: Our group working for the senior citizens is at the moment also the strongest group in the citizen initiative.
> *Frau Dörmann*: It is mostly people who aren't in work any more and have more time to get involved.
> *Frau Berghahn*: And the majority of that group were socially involved earlier too.
>
> (Group discussion, Waldstrassenviertel)

However, the form of local action and the conditions under which people could participate in them had altered in some rather significant ways since 1989. Figure 6.2, for example, shows a children's project from Plagwitz which used a focus on the environment and art to develop children's awareness of their local area. Such actions would not have been sanctioned previously, particularly as the environment had been a largely taboo subject. While work with children may apparently reinforce women's association with family and caring, the new conditions allowed women to alter the purpose and form of this work to incorporate their concerns with politics and the local environment.

In order to negotiate their involvement in this range of political actions, women also had to negotiate the range of changing conditions for their lives in the context of reunification. In particular, women sought to maintain the combination of work and family roles which had previously constituted their femininities. Even after a fall from the peak levels of unemployment in 1992, women in Saxony, the region in which Leipzig is located, still accounted for 66.7 per cent of all registered unemployed in 1995, and for 57.8 per cent of those engaged in the various job creation (ABM) and training programmes funded by the state (Freistaat Sachsen, 1996: 139). Although the educational profile of the women surveyed in the groups (50 per cent with further or higher education) suggests a relatively high status sector of the overall population is involved in these groups, attention to their employment status indicates they too had been severely affected by the processes of labour market restructuring. Although only 3 per cent

Figure 6.2 A children's arts and environment project staffed by women on the ABM programme, Plagwitz, 1992. © Fiona M. Smith

were registered unemployed, 21 per cent were on job creation programmes (*Arbeitsbeschaffungsmassnahmen*, or ABM) and 20 per cent in early retirement (available as an alternative to unemployment for those leading up to the official retirement age). In other words, 44 per cent of the women were involved in a range of measures introduced in eastern Germany by the Federal government after 1990 to cope with the extent of economic restructuring.

In addition, the nature of work for those still in employment had also changed. The change in the normal timetable of the working day meant less time was available in the afternoons for other activities and for interaction with others:

> *Frau Schulz*: You sometimes don't get home until six or half past six. In the past we were all home around three thirty. [. . .] And now almost everyone is out until the evening. And everyone has just really had enough and closes the front door.
> *Frau Walters*: And has no time for anyone else's problems.
>
> (Group discussion, Neustädter Markt)

A study by Böckmann-Schewe *et al.* (1995) suggests that in addition to the loss of work through unemployment, the work-place has also lost some of its role as a site of collectivity, social interaction and support as the organisational logic of the work-place has altered towards the efficiency principles of a market economy. Many women suggest that these twin problems affect the general willingness or ability of the population to become active in their neighbourhoods:

> People who have work at the moment have too much to do and no time to join in all these things [. . .] and those who have no work, they have no interest in it.   (Frau Lorenz, Plagwitz)

Nevertheless, there are also slippages between work and activism. Eighteen per cent of the women were members of a group because they were employed on ABM programmes to work for the various organisations. For them, the sites of activism and work coincided. In fact it was to some extent the problems of the labour market that produced an opportunity for the neighbourhood groups to apply for so many personnel to help them establish themselves and to develop particular projects. On the other hand, the limited duration of these contracts (one to two years) also created a situation of continued uncertainty for both the groups and the people employed on these contracts.

Most women continued to seek to be engaged in paid employment, retaining the identity of 'worker'. However, there was considerable disquiet with what some women saw as the narrowing of the range of femininities that were apparently required in the labour force. Frau Schwarz, for example, a middle-aged women who had previously worked as medical technician and had since had to change employment, argued that both women's ages and appearances and their desire to combine the roles of worker and mother were factors that discriminated against some women. Those not conforming to the new femininities apparently required in the labour force faced the prospect of exclusion from it:

> For me [. . .] I think now it is absolutely terrible when people now act as if there was no job left for those over 50. It is so, so depressing for them. Apparently only young, beautiful, people [are wanted] with unlimited time, and preferably people not going to have any more children.   (Frau Schwarz, Gohlis)

In particular, she argues that these new models of femininity increasingly appear to be based on women having to make a choice between career and family. She criticises this partly on the basis that it distorts women's ability to become mothers at the 'right time':

> At some point a woman must have her children. And I think they should have them at the right time and not when they are already 40 and they have to have treatment like they do over there [in the West] because the parents are too old or too worried.   (Frau Schwarz, Gohlis)

Her view of femininity ideally involved both family roles and work roles. She argued the conditions under which women are employed should not force women to choose between them. In this discussion, Frau Schwarz begins to draw out a distinctive eastern German femininity that rejects the position of women 'over there' in western Germany as either the inevitable or indeed the desirable outcome of 'transition'. This eastern German femininity is based on a continuation of the combination of mothering and full-time employment typical in the GDR:

> With us it will still be the case that for a long time, I hope, that many women, many more women will be working than over there. [. . .] It is always being prophesied that it will end up that 'with you too only just 40 per cent of women will work'. And with us, 90 per cent of us women worked and despite all the

effort required we felt good as we did it. It is rubbish when someone tries to tell us that we *had* to work. Of course we had to, but we *wanted* to work too.

(Frau Schwarz, Gohlis)

Of those women group members in paid work and on ABM schemes, the reality was that the majority continued to work full time, with some 14 per cent even working over 50 hours per week. Seventy-nine per cent also had children. In this context, the loss of childcare facilities through the restitution of buildings in which they were located to previous owners and because of the reductions in the city's expenditure threatened to disrupt women's abilities to combine these identities:

> Virtually all the buildings the kindergartens in this area are in have been reclaimed by previous owners, claims for restitution. That means that the kindergartens, the crèches have to make way. But the children are still there! And the mothers still have to go to work . . . if they still have their jobs. And we have a working group campaigning on this now.   (Frau Dörmann, Waldstrassenviertel)

> To those in the city council who say, 'oh well, in Nord-Rhein-Westfalen [a western German region] they have this and that [. . .] availability.' Well, I have to say honestly, that doesn't interest me. I want the kindergarten availability to be good *here*. [. . .] For me the kindergartens have two important functions: children learn a lot in kindergartens, like to share, to move around, to show consideration to others, to prepare for school; and at the same time they enable the mothers to go out to work without problems.

(Frau Schwarz, Gohlis, original emphasis)

On the one hand, then, these arguments establish the dual elements of work and family as key parts of women's identities. The continued desire to be able to combine them is seen as something distinctively related to their own eastern German history, with all its associated problems, but nevertheless as something that many women wish to be able still to achieve. In doing so the women draw distinctions between their identities as eastern German women and the patterns of femininity in western Germany. On the other hand, there is little sense in which women view these problems as something that might also affect men or require from them a changing masculinity. Both Frau Dörmann and Frau Schwarz argue the problem with employment changes and kindergarten closures is their effects on *women's* abilities to combine family and work. In that sense, the association of women (rather than men) with the roles of domestic care, family and parenthood remains a key element of these women's femininities.

As they seek to find ways of incorporating these different elements of their identities, some of the women involved in these groups face increased pressures which make it more difficult for them to find the time and energy to be able to exercise their desire for an active political role in the local area. However, at the same time, many women saw precisely these problems as incentives for local action. Some use these particular problems as the basis for local campaigns while others seek to connect their new-found rights as citizens with their close ties to their neighbourhoods through personal, family and caring relations to constitute new forms of local action. For the women on ABM posts, the crisis of employment restructuring actually provides support for the development of local action groups. Despite all the constraints women face, 44 per cent of the women surveyed estimated they contributed over 10 hours of work to the group each month and 24 per cent estimated a commitment of over 20 hours. Given that only 18 per cent were employed on ABM contracts, this leaves a considerable number of women involved extensively in the active contestation of the forms of local development which emerge from reunification.

Conclusion

Bütow and Stecker (1994) suggest there is a need to move away from trying to identify the differences *between* women in the 'East' and the 'West' and to see women in eastern Germany as differentiated, contradictory and complex subjects. This chapter has highlighted some of the contradictions and negotiations through which women in Leipzig sought to engage in political action, to challenge existing policies and to insist on the significance of local concerns. Women partially drew upon, and even reinforced, traditional associations of women with the local and the family in this activism, but they were also reworked through an engagement with other aspects of change. One effect was to negotiate new femininities that incorporated women's involvement in shaping change as active political subjects.

The spaces in and through which women negotiate these identities – work, home, the neighbourhood, eastern and western Germany, and so on – are all heavily gendered, in a variety of sometimes contradictory ways. Crucially, these women's activism demanded that both the neighbourhood and, on a wider scale, eastern Germany be seen as areas with distinctive concerns and as valid spaces for the development of political agendas. The women insisted on negotiating their new femininities to maintain their inclusion of the roles of 'mother' and 'worker'. In so doing, they asserted a distinctive eastern German identity, one that sometimes prioritised their position as eastern German women, and one which on other occasions drew strategic boundaries between 'East' and 'West' and sought a common cause with other people in the 'East'. The chapter has shown that women should be seen as active participants in the process of 'transition' who may even have something new or hybrid to contribute (Bhabha, 1992), and as women whose 'new' femininities are constituted in, across and between 'East' and 'West', 'past' and 'present', 'local' and 'global'. As Bütow and Stecker (1994: 324) argue: 'the image of women as active GDR-shaped subjects means eastern German women have new and different things to contribute to German unification.'

CHAPTER SEVEN

Negotiations of femininity and identity for young British Muslim women

CLAIRE DWYER

Introduction

This chapter explores the ways in which young British Muslim women negotiate competing social constructions and ideas about femininity in their daily lives. It draws upon research that focused more specifically on an exploration of the relevance and meaning of Islam for British born women of South Asian Muslim backgrounds (Dwyer, 1997). While the term 'Muslim women' risks elevating one aspect of the young women's identities (Lazreg, 1988; see also Chapter 3), I sought to consider the multiple ways in which individuals defined themselves as Muslim and the different meanings of this identification for different individuals, as well as looking at other identifications they had. The research was conducted using an interactive approach, involving group discussions with pupils in two schools in outer London. In this chapter I draw on these discussions to consider some of the ways in which the participants debated questions of femininity, particularly in the context of discussions between Muslim pupils and their non-Muslim peers. Through this discussion I hope to illustrate the ways in which intersecting discursive formations of 'race', ethnicity, gender and class are important in understanding how femininities are constructed and contested in particular spaces and places.

This chapter provides a perspective on the geography of new femininities by considering the shifting meanings of femininity for young women who are on the cusp of adolescence and young adulthood. Feminist geographers have already highlighted the changing significances and meanings of gendered identities across the life course (Katz and Monk, 1993) as well as exploring the geographies of identity amongst young people (Skelton and Valentine, 1997). Thus this chapter focuses specifically on the shifting meanings of femininity for young women at a particular stage of the life course, beginning adult life. For such young women, in the late 1990s, their identities are being shaped by many competing discourses about new femininities, from the 'Girl Power' of popular musical cultures (Spice Girls, 1997) to media reports highlighting the ascendancy of young women in education and the work-place (Wilkinson and

Siann, 1995). Such changes are also reflected in accounts that document the lives of Asian women specifically (Kassam, 1997; Bhopal, 1997).

For young British Muslim women these debates about new femininities might be considered alongside debates about the significance of 'new Muslim identities' for young people. 'New' Muslim identities reflect the reassertion of a self-consciously Islamic identity for young people (Modood, 1992). Such identities are often expressed as the embracing of 'authentic' Islamic orthodoxy which is counterposed against the 'traditional' values of the parental 'ethnic' culture. As I discuss below, these 'new' or reasserted Muslim identities may also raise questions about the renegotiation of femininity for young Muslim women (Ali, 1992). 'New Muslim identities' might be seen as one example of the 'new ethnicities' described by cultural theorists (Hall, 1992a, 1992b; Gilroy, 1993) who have emphasised the possibilities of compound or 'hybrid identities' for young British Blacks and Asians which are characterised by a mixing or syncreticism of different cultural influences. Such 'new ethnicities' have been identified particularly through more secular expressions, most notably those embodied within popular culture such as music (Back, 1996; Sharma *et al.*, 1996).

This chapter explores the meanings of femininity and the possibilities of new femininities for young British Muslim women through the 'everyday' geographies of their daily lives. It is argued that respondents negotiate their femininities in and through different spaces – in the home, at school, at work, at the youth club, community centre or night club, in the streets and shopping centre – which produce different social relations and discursive practices. The chapter examines in particular how individuals resolve the contradictions and ambivalences that such differences produce. These contradictions are discussed further in an exploration of alternative spaces of identity which are imagined by the young women – both those 'new spaces' which they might occupy in their future lives at work or at university, and those imagined spaces produced through the experimentation of identity created through the young women's experience of music, fashion or dance. Such 'spaces to dream', which might be akin to the 'paradoxical spaces' suggested by G. Rose (1993) or the 'third space' of Bhabha (1990) and Pile (1994), offer possibilities for the imagining of alternative femininities and social identities. Thus this chapter emphasises not only the spatialities of femininities but also the extent to which 'new femininities' might be engaged with through alternative imaginations of space.

The chapter begins by briefly considering the ways in which issues of femininity have been discussed in the literature about young women in general, and young Muslim women in particular. Drawing on my research findings, the ways in which the young women interviewed explore their own identities are then discussed, particularly in relation to different discursive constructions of femininity. In this section the similarities between young Muslim women and their peers are emphasised, in order to challenge some of the tendencies in dominant representations of young Muslim women to overemphasise the role of 'cultural conflict' and 'cultural differences' inscribing the lives of young Muslim women. I then consider some of the different ways in which respondents negotiate their femininity in different spaces, particularly in relation to dominant discourses about 'appropriate femininities'. Through this analysis some of the possibilities that young women have found to explore new femininities are highlighted.

Theorising about young women and femininity

There has been considerable recent research on young women's identity and experiences (see McRobbie and Nava, 1984; Weiner, 1985; McRobbie, 1991; Lees, 1993; Sharpe, 1994; Skeggs, 1997). As McRobbie (1991) has emphasised, the existing literature on youth cultures had concentrated on male-dominated subcultures and had constructed both 'youth' and 'adolescent' as masculine subjects. As the work of McRobbie (1991; see also McRobbie and Nava, 1984) and others revealed, young women were placed in contradictory positions as they negotiated competing (and often oppositional) discourses of adolescence and femininity. A key focus of this literature on young women has been the link between class and femininity – specifically the ways in which young, white, working-class women are positioned in relation to discourses about 'appropriate' femininity and sexuality (Lees, 1993; Sharpe, 1994). As Skeggs (1997) has argued recently, one of the most important considerations for such young women is the gaining and maintaining of 'respectability' in relation to appropriate femininities and (hetero)sexualities. For Skeggs (1997) it is class that is the key determinant of what alternative femininities may be available for young women.

As I have suggested earlier, the literatures about young Muslim women have often been situated outside the more general analyses of young women. Instead narratives about the lives of young Muslim women have been dominated by a 'cultural conflict' model (CRC, 1976; Watson, 1977) in which young women are defined as 'caught between two cultures' of home and school (Knott and Khokher, 1993). Through this definition, 'home' is defined as 'traditional', 'religious', 'ethnic' and is contrasted with the world of the school which is defined as 'modern', 'secular', 'Western'. These accounts can be criticised for the ways in which they pathologise Asian families. They present Asian families as the source of problems and conflicts (Parmar, 1984; Brah and Minhas, 1985) rather than incorporating an analysis of the ways in which the lives of young Muslim women are inscribed by gender relations and class structures as well as by racialised discourses. As Brah (1993: 443) argues, it is important to recognise the ways in which different social differentiations such as gender, class, ethnicity, racism and religion are 'contingent relationships with multiple determinations'. Recognising the dynamic interrelationship between structure and cultures means that while the impact of, for example, patriarchal discourses in the lives of British Muslim women can be recognised (Afshar, 1994), these are not theorised as analytically separate 'cultural constraints' but are articulated with a wider social formation (Ali, 1992). This suggests a framework for understanding how femininity is discursive, constituted within particular social, cultural and economic relations and performed through different spaces.

The research context

The research involved interviews and group discussions with 49 participants in two schools in the town of 'Hertfield' in suburban Hertfordshire.* All the participants were aged between 16 and 19. The two schools, Eastwood School and Foundation School, were characterised by some important differences. Both schools were girls' schools, although one, Eastwood School, had recently become co-educational in the lower years. Eastwood School, which might be defined as a neighbourhood comprehensive, is ethnically diverse with more than 50 per cent of pupils from 'non-white' ethnic groups, the majority of whom are from a Mirpuri Pakistani background. In contrast, Foundation School is a selective girls' school which recruits pupils from a wider catchment area. The school includes pupils from a wide variety of ethnic backgrounds with the largest minority group (30 per cent) being Jewish pupils. Muslim pupils, who make up 10 per cent, come from a variety of national backgrounds including East Africa, India and Pakistan. There were therefore obvious differences between the two schools and this was reflected in both the parental backgrounds and socio-economic status of the Muslim participants. In general, the parents of pupils at Eastwood School were employed in unskilled or semi-skilled jobs while those of pupils at Foundation School had professional or managerial occupations.

Although the research involved both interviews and group discussions as well as participant observation I draw primarily in this chapter on the group discussions. These were 'in-depth' group discussions (Burgess *et al.*, 1988a, 1988b) with the same small groups of young women conducted over six consecutive weeks. These groups were structured to allow the participants to define and develop the discussions (see Dwyer, 1997).

Debating 'appropriate' femininities

One of the most helpful aspects of the in-depth group discussions was the way in which they highlighted the social and interactive contexts within which identities are negotiated. A key focus of many of the group discussions was how the young women negotiated their identities in relation to the expectations, ideas and attitudes of others. Thus individuals emphasised the extent to which their own identities were constructed and contested in discursive contexts shaped through negotiation with parents and other family members, teachers, both Muslim and non-Muslim peers, as well as in relation to media discourses about

---

* The research was conducted in two schools in the town of Hertfield (a pseudonym) in Hertfordshire, approximately 40 miles north of London. In 1991 the census recorded Hertfield's 'non-white' population as about 10 per cent of the total (84,405). The majority of this group were defined as Asians of Pakistani origin, particularly from Azad Kashmir. Settlement by male migrants, many of whom moved from towns in the North of England, occurred in the town in the early 1970s. Family reunification took place in the late 1970s and early 1980s. Hertfield's Muslim population is therefore relatively recent compared with that in some other parts of the country. While this has some bearings on my research findings (cf. Vertovec, 1993), there are also parallels between my own research and that conducted elsewhere (Shaw, 1988; Lewis, 1994; Joly, 1995).

young Muslim women and identity. Many of these negotiations focused around what I will define as discourses of 'appropriate' femininities. In this section I want to explore some of the different ways in which these discourses were articulated and the spatialities often integral to such discourses. How the young women negotiated these discourses is explored in some detail later. While emphasising some of the similarities between the experiences of different young women in relation to these discourses, I also want to illustrate the ways in which discourses of 'appropriate' femininities are both classed and racialised.

One of the strongest ideas about 'appropriate' femininity expressed by the Muslim participants, particularly at Eastwood School, was the role of parental and familial expectations about their futures as wives and mothers. Discussing their experiences at home, individuals pointed out that, unlike their brothers, they were expected to share in domestic chores and this was seen as 'appropriate' because they were girls:

> You know you have to do the housework, girls are supposed to do the housework, help their mums. And boys can go out and . . . do whatever they like and you think . . . 'oh I wish I could do that, I wish I was a boy who could do that.'
>
> (Thaira, Eastwood School [ES])

Navida points out that it was acceptable to her parents for her to be a 'tomboy' when she was younger but as she has got older her mother has reminded her that she needs to know about domestic chores – because of her future role as a wife and mother:

> I've only recently started to do the housework, because I was quite different when I was younger, I didn't listen to my mum. Now my mum goes, 'you should do this and do that, because what are you going to do when you get married?'
>
> (Navida, ES)

If the acquisition of domestic skills was one way in which expectations about 'appropriate' femininities as future wives and mothers were articulated, another was through an evocation of the role of women as the guardians of cultural and religious integrity. In their discussions about their religious and ethnic identities many of the respondents acknowledged that it was often assumed that it was particularly important that girls knew about their religion or aspects of their culture – such as language – so that they could transmit religious and cultural values to their children:

> My mum always does that, 'You're going to have kids one day and what are they going to know about your religion?' and all this. And, 'They won't know nothing.' And maybe it's true and maybe they won't. But then it's up to me then isn't it? [laughs]
>
> (Sonia, ES)

> Actually that's what I'm afraid of because I don't really want to lose my ties. . . . What about our children and the children after? They'll totally forget, they'll totally forget that they're even Muslim. I mean if we're good mothers we'll obviously tell them. But still, if we're not good at it ourselves how are we going to teach children?
>
> (Robina, ES)

As these comments suggest, individuals reacted differently in relation to these debates. For some like Sonia, it was a matter of ambivalence, while others like Robina expressed pride in their cultural competence, such as their knowledge of Urdu, but were also concerned to know more about their religion. For many, these debates were part of wider discussions about what constituted religious knowledge, and how individuals might have access to it (see Dwyer, 1999b).

Both these discourses of 'appropriate' femininity are spatialised through a particular understanding of 'the home' as the site within which 'appropriate' feminine identities are constituted. Not only is the home the site of domestic labour, which is seen as *female* labour, but women are associated with the home as the source of cultural and religious integrity and reproduction. This spatialisation is reinforced through perhaps the most powerful discourse of 'appropriate' femininity which applied to many of the young women inter-viewed – the emphasis on young women as the guardians of family honour or *izzat* (Wilson, 1978). This concept of *izzat*, which is especially important for Mirpuri Pakistani families of rural origin, was manifest in the constant monitoring of the behaviour, and notably the attire, of young women by others, particularly when they were outside the house. As Robina explains, this often meant that the activities of young women were strictly circumscribed, if not by her own family, then by others, as they could be easily misinterpreted:

> If we go out you know someone always sees you and 'Oh God, look at her, she's out there on the street, let's go and tell her parents.' [agreement] And you get home, and before you get home, the gossip is around the whole town, you know . . . I mean even if you're not doing anything wrong . . . People are just looking for an excuse to wag their tongues about.   (Robina, ES)

This discourse of 'appropriate' femininity focused particularly upon dress. Dress emerged as the primary marker of 'appropriate' femininity for young women which was defined by the young women as an opposition between 'English' and 'Asian' clothes in which 'Asian' clothes were associated with ethnic integrity, morality and purity, while 'English' clothes were signifiers of west-ernisation, rebelliousness and active sexuality (see Dwyer, 1999a). Through this encoding of difference the young women explained how their identities could be straightforwardly 'read' from their bodies:

> If you just walk down the street and you've got trousers on, one lady says, 'I saw so and so's daughter and she's started going out with boys' . . . just because you're wearing English clothes.   (Shamin, ES)

As these quotes suggest, these discourses about 'appropriate' femininity, articu-lated through what was deemed 'appropriate' in terms of dress, were also expressed through particular spatial contexts. What might be suitable for wearing at home, or within a family party or even within an all-girls activity at the youth centre, was not suitable in the 'public' spaces of the streets which were encoded as more inherently threatening for young women both because they were outside the perceived safety of the home, and because they were seen as 'white' spaces (for a more detailed discussion of the racialisation of space, see Dwyer, 1999b). It is also clear that this discourse about 'appropriate femininity' could be mobilised in different contexts and by different people. Thus even if parents were not concerned about what the young women wore, this did not prevent others in the neighbourhood or in their wider extended family intervening, as Robina explains:

> For my Saturday job I have to wear a skirt. My parents don't mind, they under-stand you have to wear the uniform to do the job. But when my cousins came they were saying, 'What happens if they ask you to wear a low-cut blouse? Will you wear that as well?'   (Robina, ES)

Thus these discourses about 'appropriate' femininity might be understood as discourses through which the idea of community was constructed and rein-forced for members of the local Pakistani Muslim population. As I argue else-where (Dwyer, 1999b), the spatialities of a 'local, Muslim community' are made through the articulation of particular discourses of identity of which this focus upon the moral and cultural integrity of young women was one example. As Ali (1992) argues, through such discourses, which Ali characterises as patriarchal 'ethnic absolutism', a community may maintain identity in the face of perceived hostility from outsiders. As Zhora recounts below, it was also significant that it was often young men who were the most vigilant in policing young women's attire, perhaps reflecting a means by which adolescent mas-culine ethnic and religious identity could be maintained:

> I had all these split ends and I asked one of my friends to cut my hair and she trimmed it for me. When I got home my mum noticed and she had a fit. My brother was sitting there, he's a year younger than me, and he started saying, 'You'd better control this girl. She's getting out of hand. She's cut her hair, she might start wearing mini skirts tomorrow and going out with boys.'   (Zhora, ES)

As these quotes suggest, this focus on dress and behaviour, is concentrated par-ticularly on policing or 'managing' the sexuality of the young women, as Sameera explains:

> For them [i.e. their mothers] when they started their periods that meant that they had entered womanhood and they had to get married soon [agreement]. Whereas nowadays they know that girls aren't going to get married and it's just a big hassle, because what if they get mixed up with boys? . . . That's their first instinct: 'What are they going to do? Are they going to get pregnant?'
>   (Sameera, ES)

Thus the discourses of 'appropriate' femininity are crucially bound up with maintaining respectability and avoiding being labelled as someone who is sexu-ally suspect. There is an important intersection here with a broader discourse about 'appropriate' femininity for young working-class women. The strictures about behaviour and attire were far more likely to be a concern of the Muslim pupils at Eastwood School than their more middle-class Muslim peers at Foundation School, most of whom did not live in the predominant area of Asian settlement in the town. What emerged in discussions among pupils at

Eastwood School was how these discourses about sexuality and femininity were not dissimilar from those that shaped the lives of the other non-Muslim, predominantly working-class girls. Like those respondents quoted in Lees (1993), other young women at Eastwood School emphasised how important it was not to be defined as a 'slag' or as sexually available. Like their Muslim peers, such definitions were bestowed on individuals because of their clothing or perceived behaviour and were also spatialised – being out on the streets at the wrong time, late at night for example, could mean risking such a reputation. There are, therefore, some important parallels between Muslim pupils and their non-Muslim peers about discourses of 'appropriate' femininity, particularly in the ways in which it concerns sexuality.

Before going on to consider how the participants interviewed in both schools responded to these discourses of 'appropriate' femininity, I want to look at some of the other ideas about femininity that shaped their discussions. If ideas about femininity were transmitted to the young women through their parents, they were also gained through discussions with teachers and peers at school. By focusing on the school, however, I want to resist the home/school dichotomy that has structured many accounts of the lives of young British Muslim women. As Knott and Khokher (1993) argue, such a dichotomy both reinforces 'culturalist' explanations and fails to understand the complex ways in which religious and ethnic identities may be being constructed. It is clear that at school individuals came across ideas about 'appropriate' femininities which both reinforced and contested the ideas gained from other sources.

In their discussions pupils at Eastwood School agreed that, while the school generally did much to help Muslim pupils, teachers' underlying expectations of Muslim girls often reflected a particular view of Muslim femininity:

> Also a lot of the girls, they were all Asians so after school they would get married and have kids and that would be it and none of them went out and got careers and stuff like that. So it was sort of like a 'no-hope school', where people just got married.
> (Zakiya, ES)

These expectations were also confirmed in discussions that I had with teachers at Eastwood School. Many of them saw the time spent in the sixth form (final years of school) by Muslim pupils simply as a postponement of their inevitable marriage. These expectations are classed as well as racialised, since expectations for all pupils at Eastwood School were not high. In contrast pupils at Foundation School emphasised that the school articulated a strong discourse of 'equal opportunities' and pupils were expected to set themselves very high standards. Indeed, there was even a concern among some pupils that to aspire to the roles of wife and mother was not acceptable, as one orthodox Jewish pupil admitted:

> Don't get me wrong, I want to have a career, and, I am ambitious, but . . . it's very important for me to have a family, you know, to settle down, get married and have a family and then the career to me is second. Now you may say that that is wrong, but to me, it is so much more important to have a family.
> (Hayley, Foundation School [FS])

Most of the young women at Foundation School, both Muslim and non-Muslim, emphasised their intentions to go to university and pursue professional careers, although this did not mean they ruled out motherhood in the (long-term) future. Yet even at Foundation School pupils often came up against stereotypical views of 'appropriate' femininity for Muslim girls, as one respondent remembers:

> When I had my [mock university] interview the first thing the doctor asked me was 'I know that a lot of Asian girls want to go to Medical School, and they get into school and later on their parents give them hassle, what have you done about that?' And I, I don't know, I mean, I said that they are not all like that, you know, I've got very understanding parents, I've talked to them about how it's a six year long degree and at the end of it you have to work weekends and late nights, with me, you're treating men, and they've accepted that. I said, 'we're not all like that, you know.'   (Riffat, FS)

Thus at both schools pupils were confronted with ideals about femininity that were most strongly inflected by the dominant class positions of pupils at each school. Yet at both Eastwood School and Foundation School, as I have suggested, these expectations were also racialised through assumptions about appropriate femininities for Muslim girls. These expectations were also reinforced through stereotypical attitudes about Muslim women that individuals encountered from both their peers and some teachers:

> All the white girls think that all Asian girls have arranged marriages and all these Asians are sent to Pakistan. Like if an Asian girl goes to Pakistan, even for a holiday, they say, 'Oh she's gone to get married.'   (Sonia, ES)

> Other people think Muslim women are not allowed to go out, they are not allowed to do this, they have to cover themselves, they're chained to the kitchen sink. But we're not like that.   (Robina, ES)

Many of the respondents emphasised that such stereotypes were promoted by the media's portrayal of Asian and Muslim women (Parmar, 1984). A particular example of this was a television series called 'Bounty Hunter' which sensationalised stories about young Asian women fleeing from their families to avoid arranged marriages. Participants argued that such stories reflected particular circumstances and contexts – for example, they argued that the stories drew upon the experiences of young women from particular class backgrounds – yet for their non-Muslim peers these programmes simply reinforced stereotypes about a docile Muslim femininity and tyrannical parents:

> You know the stereotype, you've got the little timid Muslim girl with the headscarf on, and the timid mother, and the strict father or grandfather who won't let her go out or who won't let her do anything.   (Sara, FS)

Negotiating 'appropriate' femininities

Having outlined some of the different ways in which discursive constructions of femininity for Muslim girls were debated by the respondents, I now consider how the girls negotiated these ideas within their daily lives. In this section I consider the 'everyday' negotiations of competing discourses of 'appropriate' femininity described by the respondents and in the next section I look at some of the alternative femininities that they explore. One of the strongest points that emerged from discussions with participants was the extent to which many of them felt that they were continually resisting other peoples' assumptions about them. These resistances could also appear to be somewhat contradictory. Thus individuals might resist parental prohibitions at home about what 'a good Muslim girl' should do, while simultaneously, among peers at school, they would challenge stereotypical ideas about Muslim girls being oppressed by their parents.

As outlined above, clothes were often used as the means by which femininities were expressed. For many of the respondents, negotiations of their femininity meant challenging the meanings which others attached to the clothes that they were wearing. In particular, many individuals sought to challenge the meanings attached to an imagined dichotomy between 'western' and 'Asian' clothes (for a fuller discussion of these ideas, see Dwyer, 1999a). In relation to the parental discourses that identified 'appropriate' femininity with wearing 'Asian' clothes, participants emphasised that there were no inherent meanings associated with particular forms of dress:

> It's like wearing a long skirt, wearing westernised clothes, which cover you up. We say we're right because we're covering ourselves and there's nothing wrong with wearing it. . . . It's like the other day I came in, and the skirt I had on was right down to my ankles. My uncle was sitting there, and my mum said, 'Go and change your clothes.' And I said, 'No, I'm not going to change my clothes.' And I was right to take a stand because I knew I was right. (Sameera, ES)

> Like it doesn't say in the religion . . . you've got to wear Asian clothes or anything, it just says that you've got to be covered. (Ghazala, ES)

While the participants contested the idea that 'Asian' clothes had any inherent meaning, they also challenged the assumption that any clothing could necessarily define your sexuality and femininity. As Humaria suggests:

> There's this girl we know and like she . . . she's quite all right, like one of us and that . . . but . . . the only thing about her is that she goes out late at night with her brother or her friend and she dresses in short skirts and stuff. Now she could probably be really trustworthy and her parents could rely on her for anything and she probably doesn't go out with boys or anything but the fact that she . . . wears short skirts and stuff . . . gives her a really bad reputation.
>
> (Humaria, ES)

Similarly, conforming to 'appropriate' dress codes did not necessarily mean that the young women were conforming to 'appropriate' feminine behaviour, as Ghazala argues in relation to the most contested piece of 'appropriate' feminine attire, the headscarf:

> If there is a girl with a scarf on her head, right, and she's been out with all these
> guys, she'll get away with it because she's got that cover, it doesn't matter how
> bad she is.   (Ghazala, ES)

Yet if individuals challenged the meanings about femininity that were attached
to dress by their parents, they also confounded the assumptions of their peers,
that just because they chose to wear 'western' clothes they were necessarily
'rebellious' or not interested in their religious or cultural heritage. As Ghazala
explains:

> It's like this girl, she was saying, 'you're modern, westernised people. Why are
> you hanging round with . . . like, people with scarves on their heads and that?'
> She didn't expect it. But they're my friends . . . just because I wear jeans and a
> leather jacket it doesn't mean I'm a rebel or something.   (Ghazala, ES)

These challenges to the femininities that were imaginatively attached to different forms of dress were also negotiated by the respondents in relation to
different places. As suggested earlier, in many accounts of the lives of young
British Muslim women there is a tendency to dichotomise the space of the school,
understood as secular, permissive and 'free', with the space of the home, represented as religious, repressive and restricted (Knott and Khokher, 1993: 595).
Yet pupils' own understandings of the ways in which they might express their
feminine identities, through dress, within these spaces did not necessarily
reflect this simple binary. It is true that some individuals experimented with
their appearance in different spaces. For example, walking to school they wore
headscarves to articulate a 'safe' and 'appropriate' Muslim femininity. Once
at school, however, headscarves were removed and, like schoolgirls everywhere,
they experimented with new hairstyles and make-up, before donning scarves
again to return home. Yet for other pupils the opposite might be true. While
at school they conformed and wore a school uniform that others might assume
reflected 'Western' femininity, they felt more comfortable experimenting with
Asian styles of dress, which were deemed both more varied and more attractive,
when they were at home:

> No, no . . . with English clothes, they're sensible for school and stuff like that,
> but personally I prefer Asian clothes. Because we've got more, I don't want to
> be horrible about English clothes . . . but we've got much more variety. With English
> clothes they've got two sorts, skirt and top, trousers, that's it. But with Asian clothes
> you've got saris, *shalwar kameez*, a long skirt and short top.   (Sarah, ES)

> If you're in your own clothes, you get the feeling, 'Oh, this is me'. Like when I
> wear my uniform I think, 'Oh, I'm a pupil now'. You know it is, like it represents
> you . . . but when I put my own clothes on, I think that this is me now, this is
> the real me. I feel much more comfortable in my own clothes, because not just
> that they're trendy, but they are so relaxing and they're open.   (Eram, ES)

As I have suggested already, the home/school dichotomy has been critiqued
because of the ways in which it is simplifies the cultural and religious identities of individuals, and leads to 'culturalist' explanations. This criticism is

evident in the analysis of what might be deemed 'alternative' Muslim femininities, considered in the next section.

Constructing alternative Muslim femininities

As I will argue in the concluding section of this chapter, the young women interviewed were all involved in different ways in experimenting with the construction of alternative femininities or the negotiation of 'appropriate' femininities. However, perhaps one of the most important alternatives being debated by the participants was the adoption of a more self-conscious Islamic or orthodox Muslim identity. Such 'new' Muslim identities (Modood, 1992; Samad, 1993) have been widely debated, particularly as to whether they represent self-consciously 'religious' identities (Knott and Khokher, 1993; Jacobson, 1997) or are rather a mobilisation of 'residual' Muslim identities (Lewis, 1994: 177). This distinction about what exactly the evocation of an explicitly *Muslim* identity might mean is considered in more detail in Dwyer (1999b). What I want to highlight in this chapter is how discourses about 'orthodox' Islam could be mobilised by pupils to construct alternative femininities.

It was evident that in both schools the meanings of 'Muslimness' were being debated by pupils, reflecting both wider globalised discourses of Muslim identity (McLoughlin, 1996; Werbner, 1996), as well as the aftermath of the so-called 'Rushdie Affair' within the British context (Appignanesi and Maitland, 1989; Rutven, 1990). The adoption of a more explicitly Muslim identity by some pupils was reflected in the rejuvenation of Muslim societies within each of the schools at which visiting speakers – often from organisations such as Young Muslims UK (for an analysis of this and other Muslim organisations see Lewis, 1994: 102–112) – prompted debates about what it meant to be a Muslim. Such 'new' Muslim identities were characterised by a search for 'orthodox' Islam. This version of Islam was often counterposed to views of Islam transmitted by parents which were rejected as being more about 'Pakistani culture' than religion. Adherents to a 'new' Muslim identity sought authority through a close reading of the Koran, rejecting the oral teachings of parents or the authority of Pakistani clerics (*maulvis*). An important emphasis here was the rejection by the young women of parental prohibitions about 'appropriate' feminine behaviour or attire, which were regarded as being the result of cultural prejudices rather than having any rooting in orthodox religious teaching. This view is exemplified by Ruhi:

> This is something I feel so strongly about. . . . People say that women are oppressed in Islam. It is 100 per cent true, women are oppressed, but it's not Islam that oppresses them. It's the culture that oppresses Muslim women. It's the men that oppress Muslim women. It's society that oppresses Muslim women. Not Islam. If I was to go out there and, you know, take all the rights that I have, the rights given to me . . . I would be the talk of the town. People would say, 'Oh look at her, she's getting well out of line.' But I know that Allah has given me this right, so why shouldn't I fulfil it?   (Ruhi, ES)

As Ruhi suggests, the articulation of a 'new' Muslim identity was particularly attractive to young women because through it an alternative Muslim femininity could be constructed. Through this discourse young Muslim women gained greater freedoms but did not compromise their integrity in relation to religious values. In other words, religious authority could be mobilised to support their claims which then carried a greater moral superiority in their arguments with parents. This alternative femininity was expressed most powerfully in the adoption of an explicitly Islamic dress code. In contrast to the *shalwar kameez* and loose headscarf usually worn by women from the Indian subcontinent, this Islamic dress usually took the form of either long skirts or trousers and a more complete, 'Middle-Eastern style' head covering defined as '*hijab*' (Ali, 1992: 114). This form of dress expressly challenged the idea of 'Asian' clothes as boundary markers for cultural and religious identity, while also making a firm statement about an explicitly religious identity.

This adoption of an alternative Muslim femininity raises some interesting contradictions. First, as individuals such as Ruhi asserted, it was an extremely powerful means by which young women could challenge the authority of their parents from within the boundaries of discourses about 'appropriate' femininity. It became a means by which parental ideas about 'appropriate' femininities could in effect be challenged and subverted. As Ali (1992: 114) argues: 'the good Muslim girl who shows unusual devotion of her faith may find it possible to express a desire for higher education or professional employment without risking her position or that of her family.'

Thus the adoption of an explicitly Islamic identity, symbolised in particular by the wearing of the *hijab*, could be a means of resistance for some individuals. However, as Alibhai-Brown (1994) implies, this alternative Muslim femininity – particularly expressed through the wearing of the *hijab* – raises some contradictions. The wearing of the veil, while seen by such individuals as a powerful statement of their own independence, remains complicit with a 'rhetoric of the veil' (Abu Odeh, 1993), which scripts feminine sexuality as dangerous or threatening, such that it must be constrained. One way in which this contradiction was played out among my participants was the resistance by some pupils to the wearing of the *hijab* by their fellow Muslims. Pupils recognised that the meanings attached to the veil, associations of 'purity' or 'morality' (Ahmed, 1992), meant that those who chose not to wear it were defined in opposition to these meanings. As one pupil wryly remarks:

If you get the wrong ideas inside you. Say you think about sex 24 hours a day [laughter] with a scarf on your head. Then why wear the scarf? It's just a cover up.
(Husbana, ES)

Thus alternative Muslim femininities appear to be contradictory because they remain complicit with dominant discourses about feminine sexuality. It might be debated just how far it is possible to construct a new Muslim femininity against the powerful associations attached to the veil (see the further discussion in Dwyer, 1999a). However, at the same time, such alternative femininities did offer possibilities for the participants to construct alternative identities – and futures.

The expression of these alternative Muslim femininities, particularly through dress, are also articulated through different spaces. One way in which this occurs is that the wearing of the *hijab* can be a 'safe' means by which unfamiliar or public spaces can be negotiated without fear. As I have suggested earlier, individuals might wear a headscarf walking to and from school then remove it at school. Similarly, several participants who did not yet wear the *hijab* talked about possibly adopting it when they went away to university, suggesting that it would make them feel more comfortable in an unfamiliar and perhaps threatening environment, as well as allaying parental fears:

> Probably if I go to university it's a fresh start there and I might actually consider wearing a headscarf, because no one will know me there, and you feel more secure.   (Sara, FS)

Thus the wearing of the headscarf becomes a means by which new public spaces can be negotiated safely. (There is an interesting parallel here with the work of MacLeod, 1992, who describes how 're-veiling' has become an 'acceptable' means by which lower middle-class women in Egypt can enter the 'new' spaces of work opened up through processes of modernisation and social change.) The spatialities of identity were also emphasised in a different way by Ruhi, one of the participants at Eastwood School who was most proactive in the adoption of an alternative Muslim femininity. Ruhi had faced considerable opposition from her parents, particularly her mother, to her wearing of the *hijab* at home. Instead school had become the place where Ruhi had been able to express her 'new' Muslim identity more confidently. This suggests that while school might be a site for the experimentation of identity for young women, the resultant identities are not necessarily 'Western' and secular ones.

Negotiations of femininity and identity: imagining futures

Having explored debates about 'appropriate' femininities, and the possibilities of alternative Muslim femininities, in this final section I want to examine both how the participants experimented with and explored different femininities and how they imagined their future lives. As was suggested at the outset, the participants of this research are all at a particular stage in the life course, at the cusp of adulthood, a period characterised by the exploration of their adolescent femininity and sexuality. These young Muslim women's identities are overdetermined by dominant discourses about 'appropriate' Muslim femininities which come both from parents and from the stereotypical ideas expressed by teachers, the media or non-Muslim peers. Against this background individuals sought to challenge the meanings attached to them in particular places or to seek alternative meanings. One of the most interesting ways in which such challenges were articulated was through what might be characterised as alternative imaginative spaces of identity. A good example of this is a fashion show organised by pupils at Foundation School.

As the participants explained to me, the aim of the fashion show – entitled 'East meets West' – was to challenge dominant perceptions in the school about Muslim girls and to offer an alternative to the 'Asian'/'British' dichotomy through which femininities could be constructed. The show was intended to demonstrate self-consciously the possibilities of 'hybrid' identities through the blending and crossing over of 'British' and 'Asian' styles. As one of the organisers explained:

> We, and some others, are mixing. Like this sort of skirt [indicates Aisha's long silk skirt], this is really nice. The skirt is sort of Asian colours and the material. And the top that goes with it is Asian, and the scarf.   (Riffat, FS)

Models in the show wore clothes that were deliberately chosen to confound an opposition between 'Western' or 'Asian' fashion and to emphasise the ways in which fashions cross these boundaries. The choreography of the show was also designed to emphasise this point (but see Dwyer, 1999a). The fashion show was a very self-conscious negotiation of identity by individuals to challenge other people's perception of them (Bhachu, 1993: 111). It was also an event that can be read as the opening up of an alternative imagined space – outside the bounded dichotomies that overdetermine the identities of young Muslim women – through which new forms of identification and femininity could be imagined. Such imagined spaces might parallel the 'hybrid' musical spaces suggested by Back (1996) in which 'new ethnicities', new forms of ethnic identity which subvert older binaries of black/white or Asian/British, can be performed. In the case of the fashion show, it also provided an alternative imagined space for young women to experiment and 'play with' their feminine identities. As was clear in the fashion show performance, participants gained obvious enjoyment from experimenting with different styles and femininities within the temporary space opened up on the catwalk.

A rather more mundane example of this experimentation was also evident at Eastwood School. On several occasions after meetings of the Muslim Society I observed young women in the girls' toilets (an important symbolic space for the exploration of feminine adolescent identities) experimenting with wearing the headscarf and practising different styles of securing the veil. Here too young women were involved in 'trying out' an alternative feminine identity within the 'safety' of a separate space. Significantly, as Abu Odeh (1993: 34) also remarks, young women were exploring the ways in which wearing the headscarf gave them a confidence in their feminine sexuality, confounding some of the assumptions often made about the veiled woman. Both of these examples provide evidence for the ways in which the respondents were involved in experimenting with and 'trying out' alternative feminine identities. Both the fashion show and the girls' toilets become alternative spaces within which 'new identities' can be both imagined and performed beyond the bounded spatialities, with their overdetermined meanings, of the young women's everyday lives.

If participants were able to explore alternative femininities through these imagined spaces, they also talked about their identities in terms of imagined futures. In response to some of the discourses of 'appropriate' femininities discussed earlier, the respondents considered how they might hope to shape their own future lives. Overwhelmingly, participants saw their futures in terms of wanting 'independence', particularly in terms of education and work, before choosing to get married or have children:

> I don't want to get married so young. I just feel that I'm not ready and some Asian girls who get married when they are just 17 or 18 or 19 they haven't lived yet, you know. I want to be at least 20 or 24, once you've had all you want, all your life.
> (Robina, ES)

> I want a career before I get married . . . I'd prefer to get married at a lot later age . . . than early . . . because most women . . . Muslim girls get married about 20 to 22, arranged marriage . . . I'd prefer to have a career. I'll probably go to university if I pass my [exams] . . . and have a career.
> (Thaira, ES)

The ways in which these respondents talk about the future suggest that marriage and having children, 'an appropriate Muslim femininity', is viewed as an ending point. Almost, as Robina suggests, something that happens once they have had 'all your life'. Such a perspective is not perhaps surprising when interviewing any young women in their late teenage years. What I want to highlight, however, is the ways in which the respondents negotiate these expectations of marriage and motherhood, by carving out a space that will be their own, at university or at work, before they will conform to expectations. This transition period, which might be understood as negotiating additional time and space, marks the means by which young Muslim women are involved in gradually altering the models of 'appropriate' femininity that might be available to them. Indeed several respondents contrasted their expectations with those of their older sisters. It was evident during the research that a number of factors helped in the negotiation of this transition period. Socio-economic changes meant that levels of unemployment were high amongst unskilled Asian men in Hertfield while it was recognised there were more jobs for young women, particularly in the service industry. In these circumstances working-class families were more likely to see that work was necessary for their daughters to contribute to the family income. It was also evident that mothers were generally very supportive of their daughters' aspirations, often contrasting the opportunities they saw for their daughters with their own lack of education:

> My mum says, 'when you're older don't do what I did. Go to university . . . get a job and be independent. Because you know the world is changing and no one is going to be there for you forever.'
> (Habiba, ES)

> My mum didn't have the chance to have a good education and she's trying to educate all her children, to, you know, to get somewhere. That's her main ambition really, to have all her children educated really and working and that. And that was a shock to me, because she didn't really have that much education,

and I thought she might want me to stay at home and get married at the age of 16 or whatever but she doesn't.   (Husbana, ES)

Such findings echo Afshar's (1989b) three-generational study of South Asian Muslim women in Bradford, in which she argued that young women received considerable support from their mothers to use education to develop an alternative life path and feminine identity.

Conclusion

In this chapter I have explored some of the ways in which the young British Muslim women interviewed negotiate their identities. I have emphasised the extent to which such negotiations are made in relation to the expectations about Muslim women that are held by others, both 'community' or family discourses of 'appropriate Muslim femininity' and the stereotypes about Muslim women that are produced by the media and reproduced in many different social or educational experiences. My intention has not been to deny the significance of these discursive constructions and the real effects they have in the lives of young women. Instead I have chosen an analysis that reflects the multiplicity of subject positions which might be held by young Muslim women while also recognising the ways in which their everyday lives, like those of all individuals, are constituted through intersecting discursive, material and social formations. One example of this is to emphasise that the different class positions occupied by different individuals, reflected here particularly when contrasting the experience of pupils in the two schools, were important in shaping both ideas about 'appropriate' femininity and the material and social conditions of the lives of individuals.

In conclusion, I wish to emphasise the geographies through which the young women represented here construct, contest and negotiate their femininities. I have emphasised the ways in which discourses about 'appropriate' femininities for young Muslim women are often articulated through ideas about 'appropriate spaces'. Thus the home is the site that embodies both female labour and feminised cultural values of ethnic purity and integrity. Spaces outside the home – particularly the 'public space' of the street – are encoded as less secure, requiring young women to exercise greater vigilance in assuming 'appropriate' femininities. I have also suggested some of the different ways in which the young women interviewed respond to these ideas and negotiate their identities differently within different spaces, whether this includes conforming to particular assumptions within 'public spaces'; adopting alternative Muslim femininities to explore 'new' or potentially threatening spaces; or using the 'safe' space of the school to experiment with multiple feminine identities and to imagine 'alternative spaces'. In exploring these ideas I have tried to move beyond a simple binary opposition between 'home' and 'school', suggesting that the alternative femininities explored at school may not necessarily be those that support 'Western' or 'secular' values. Similarly, I have emphasised that 'home' is not simply a site of oppressive patriarchal values but is also a space within which young women can also explore alternative femininities outside assumptions about the meanings of cultural difference. This chapter has illustrated how some young British Muslim women at a particular point in their lives, poised between adolescence and adulthood, engage with (and sometimes transform) the range of feminine identities available to them as they embark upon their futures.

CHAPTER EIGHT

Configuring and reconfiguring geographies

Introduction

The preceding four case studies have illustrated diverse circumstances within which different women engage with the everyday expressions of social change. Their responses, which often require considerable effort and energy in difficult circumstances, produce a multiplicity of 'new' femininities. The femininities discussed in these chapters are constituted in part in relation to a range of processes which we identified in Chapter 2 as key elements of the large-scale economic, cultural and political changes involved in globalisation, the spread of neo-liberalism and processes of development and post-communist transition. Thus new femininities are constituted in relation to changes in employment, access to resources, provision of public and private services, as well as the circulation of people through migration. Furthermore, the case studies indicate the importance for new femininities of the often hybrid processes involved in the circulation of cultural, political and religious ideas, ideologies and practices. Each case study also deals explicitly with the multiple constitution of femininities which was identified in Chapter 2 as a significant theme in thinking about changing gendered identities.

In this chapter we draw together all four case studies to explore the implications of our findings for understanding the geographies of changing femininities. We begin by examining the importance of geographies in configuring new femininities in the context of the ideas about global/local change and fractured identities mentioned above. We work through the typology of ways in which geographies are involved in the constitution of femininities, which we introduced in Chapter 1. In the second section of the chapter we consider the implications of these arguments for our understandings of change, asking firstly how a *gendered* perspective might require a reconsideration of the theories of social change and, secondly, how a *geographical* perspective can illuminate different understandings of how femininities are reworked. These discussions are then developed further in Chapter 9, where we deal more explicitly with the diverse and sometimes conflicting ways in which these changing femininities may in fact be 'new'.

Geographical constitution of femininities

In Chapter 1 we suggested that geographies are important in the constitution of femininities in three particular ways: in the difference that place makes, in the everyday spaces of life, and in the gendered understandings of these spaces. We now revisit and develop this typology in the context of the case-study chapters, and with respect to the issues of global change and the need to deal with difference highlighted in Chapter 2. Firstly, we argue that place does indeed make a difference: the different positions of Britain, eastern Germany and Peru in the world economy and political order, and their different national and sub-national histories, economies, cultures and political systems, matter in the constitution and contestation of new femininities. We develop this argument by stressing the shifting interconnections of the global, the national and the local, exploring how places are at one and the same time distinct and yet intensely bound up in other networks and processes. Following this, we argue that gender identities are not fixed social categories which are either universal or entirely discrete, but rather they must be seen as relational and situated in the interconnections of processes operating across place and scale.

Secondly, we contend that the different sites and networks of everyday life are important, as they offer different potentialities for the (re)construction of gendered identities. Thus, for example, femininities may be shaped and performed in different ways in the paid work-place compared with the home. Moreover, local social networks can act as effective moral communities sanctioning some social practices but not others, and may also offer possibilities for individual and collective social action. This emphasis on the importance of these everyday spaces highlights a third, and dialectically related, way in which geographies are important in the constitution of femininities. Our understandings of particular spaces or sites are themselves actively produced through discourses about gender; thus particular spaces, such as those in the public sphere, are coded as male, and others, such as the home, as female. These understandings are often complex and may well be racialised and sexualised (and so on), as well as gendered. Below, we draw on the case studies to illustrate this dialectical relation between gender and space in the section entitled 'everyday spaces'. First, however, we turn our attention to the difference that place(s) make(s).

The difference that place(s) make(s)

Perhaps the most obvious way in which 'geography' is important in our discussions is that the case studies are located in differing parts of the world where the intersections of past and current developments provide specific contexts for the constitution of new femininities. These 'places' can be located at a series of scales, from the international, through the state, to the intra-state region. Most broadly, at the international scale, Britain, eastern Germany and Peru all occupy different positions in the global economy and political order, as well as having different national histories, economies, cultures and political systems. Britain's status as a world economic and political power has declined rapidly during this century, but the remnants of the past and its growing links within the EU mean that it still clings to its status as a 'First World' nation. It remains an important player among western democracies through involvement in such international organisations as the economically powerful 'G8' group of nations and the military alliance of NATO. Domestically, Britain's relatively powerful position in the world has allowed it to enjoy extensive formalised structures of employment and wealth creation and, although weakened by neo-liberal agendas in the 1980s in particular, a significant welfare state system.

Eastern Germany has a very different, and changing, position within global structures. Having for 40 years been a leading 'Second World' nation on the 'front line' of the Cold War, it has experienced considerable and rapid change since the demise of the political and economic alliances of the Warsaw Pact and COMECON. Reunification with West Germany and related membership of a range of international alliances, including the EU, has brought accession to a pre-existing system of western liberal democracy. This has allowed many women (though by no means all) new possibilities for democratic action and for access to goods and services, travel, technology and forms of employment, education and culture which were not available in the past. However, the extension of the free market has also caused a rise in unemployment and a restructuring of existing employment in line with the introduction of market relations. This has been paralleled by cuts in state provision for workers and families. Reunification also gives eastern Germany a position that is distinct in several key ways from other former communist states where struggling economies and ethnic conflicts present more severe problems.

Peru's status as a 'Third World' nation was arguably strongly influenced (some would say, caused) by the manner of its incorporation into the global economy when it came under Spanish colonial control in the 1500s. The effects of this colonialism have been long lasting and they continue in current neo-colonial relations between Peru and powerful political and economic interests in the 'first world' and among international institutions (Barton, 1997). In the 1980s, Peru, like much of Latin America, was hit by economic crisis which was worsened by the withdrawal of international financial support when the president of the time, Alan García, curbed debt repayments. President Fujimori's rise to power in the 1990s then brought a series of stringent structural adjustment policies – liberalising the economy, privatising state industries, reducing state welfare provision – aimed at renewing international financial confidence in the country but which impoverished many of its people. The distinctive positions of these countries in the global economic and political order clearly influence the life chances of their citizens, facing them with different opportunities and difficulties.

Significant to our argument is the fact that these global processes are gendered, shaping women's and men's lives in different ways. The PAIT programme in Peru, introduced as a response to the processes described above, demonstrates this point well. The emergency work programme was aimed at men, thus the policy was overtly rather than covertly gendered; nevertheless, the results were contradictory as more women were drawn into the work-force.

Similarly, the massive economic and political restructuring of eastern Germany has had gendered effects with unemployment among women being a particular problem. Equally, the women in the Sheffield case study who were drawn into the work-force are part of the feminisation of labour that has been part of economic restructuring in Britain. Economic restructuring has also had a disproportionate impact on ethnic minority groups in the UK, often affecting men more than women. These changes have been significant in shaping the gender relations governing the lives of the young Muslim women in Dwyer's study. It is not only in economic terms that these processes are differentially gendered. The rural/urban migration affecting women in Peru, for example, or the migration processes within systems of colonialism and post-colonial relations that structured the lives of the relatives of the young Muslim women in London also operate in ways that selectively alter the life experiences of women and men. Migration to the UK opened up new opportunities in the lives of some women, while limiting others. Our point, then, is that place in the world matters in the constitution of gendered identities as the processes shaping the global order do not shape generic human subjects. They shape gendered men and women in relation to one another and to women and men in other parts of the world. Gender identities and one's (changing) place in the world order are intimately related.

The different countries, which the processes of globalisation permeate, are themselves characterised by different gender regimes, that is, dominant ideologies of gender identities that are underpinned (though not completely, and not without contradictions) by and through the policies and institutions of the state. These may operate in tandem with other social institutions or religious organisations, for example. These gender regimes may of course be altered by a range of processes such as political action, by the effects and demands of economic and cultural change, or by the very contradictions that are opened up by the actions taken by individual women and groups of women (and men) in shaping their daily lives. Nevertheless, the dominance of these regimes means they are important factors in defining the possibilities different women have in constituting new femininities. This is perhaps best illustrated by some examples.

During the course of the twentieth century women's legal subservience to men has been gradually removed in Britain, as it has in many other countries, as women have gained access to the vote, 'equal' pay and employment rights and so on through a combination of individual choices, political action and social and economic necessity. Alongside this rapid and widespread change, gender differentiation organised around the heterosexual model of male breadwinner and female housewife has remained important in Britain, both socially and politically. However, economic changes have made such household arrangements much harder to sustain, for both the 'white' families in Sheffield and the British Muslim families in London which are considered in Chapters 5 and 7. The contradictions between different facets of the gender regime (such as economic necessity and social welfare models) create difficulties and opportunities, spaces of tension in which individuals, families and the state, for example, produce different responses.

A similar set of tensions was apparent in communist East Germany where women and men were officially designated as equal citizens, but where the state patriarchal system also assigned women the primary responsibility for domestic issues and the reproduction of the family. Differences between women and men in their assumed responsibilities are now being exaggerated and reinforced in selective ways in the period after reunification. Dominant western German discourses about gender already shape national frameworks of taxation, welfare and employment, leaving eastern Germans to work with the implications of these 'western' discourses for the specific conditions of the eastern German regional context. In this period of change some revisions to abortion law represent one of the few areas in which eastern German femininities have been incorporated (albeit highly problematically) into the new national gender regime (Riemer, 1992; Maleck-Lewy, 1995). The coming together of the western German regime with reactions against, and the experiences of, past eastern German gender regimes presents a complex and highly contradictory context within which eastern German femininities are being reworked (see Schenk, 1995).

In Peru the state has expended considerable effort over a long period promoting domestic femininities for women, popularising in particular the idea of the self-sacrificing mother. This domestic gender regime is in turn underpinned by transnational gender ideologies of *marianismo*, as well as by the funding strategies of international NGOs which have supported women's activities in soup kitchens and related survival strategies since the early 1980s. Although it could be argued that these domestic ideologies ensure social stability, they have also been appropriated by women wishing to challenge the state in times of economic and political crisis. Consequently, the soup kitchen movements and the 'glass of milk a day' scheme have provided spaces for leadership training for women (Barrig, 1989; Vargas, 1991) and for opposition towards the activities of both the government and the Shining Path (*Sendero Luminoso*) during the civil war (1980s to early 1990s). The political importance of these movements can be seen in the fact that the leader of the 'glass of milk a day' scheme, Maria Elena Moyano, was assassinated in a particularly horrifying way in 1992 by the Shining Path for refusing to join their forces and sever her relations with women's grass-roots networks. According to DESCO, one of the best known research institutes in Peru, this was because:

> she represented a way to be a leader and a woman that was alternative to the paradigm of a woman of war portrayed by Sendero. Her image affirmed the rights of the citizens, the competence of the government and the important role of women in constructing peace. It therefore contained a message that questioned the value that the Sendero armoury recognise in women. (Mavila León, 1992: 49)

Thus Peruvian gender regimes based on domesticity have played an important role in the struggle for an alternative to civil war. The investment of the state or other powerful groupings in these gender regimes may be great enough to place women who practise non-conforming femininities in dangerous positions.

The example from Peru also suggests that however important the state may be in shaping gender regimes, other powerful gender ideologies exist outside the state. As Chapters 2 and 6 argue, the media and global patterns of consumption remain important forces for the circulation of a variety of representations of and possibilities for femininity. However, their impact is far from clear cut. Similarly, gender ideologies may transcend state boundaries through migrations and the creation of diasporic populations. The interpretations of 'appropriate' femininities which the young women negotiate in diaspora Pakistani and Muslim contexts alongside a hegemonic 'British' gender regime suggest the existence and importance of what we might well call transnational gender regimes, constituted through institutions and discourses other than the state. Nevertheless, while both state-led and transnational gender regimes are important, they are neither exclusive of other influences nor without their contradictions. These varied and often intersecting gender regimes remain important as they provide different contexts, constraints and possibilities for the constitution of 'new' femininities, but they cannot be considered as fixed entities with impervious boundaries, nor are they exclusive of other influences that may well disrupt their power.

Considerations of macro-regional differences and state-level distinctions (and the relationships between them) do not fully cover the effect of place on the development of femininities and neither they, nor their gendered effects, are without their contradictions and inconsistencies. This is also true when we move to the intra-state scale. The case-study areas are located within particular regions which have, through their history and contemporary circumstances, been drawn into global structures in different ways and are characterised by dominant forms of gender relations which are constituted both locally and in relation to wider processes. For example, Sheffield, one of the British case studies, has, because of its heavy industrial history, been hit disproportionately hard by the shift away from manufacturing towards the service sector and high-technology industry. The response to the deindustrialisation in Sheffield in terms of gender has been different from that in some other areas faced with similar problems. In Sheffield women have been involved historically in the cutlery trade but, unlike male employment, this involvement was seldom politicised, and in the process of deindustrialisation it was male unemployment that constituted the focus of local concern. Such concerns have been popularised through the film 'The Full Monty' (1997) which concentrated on the crisis of masculinity caused by unemployment and recession, rather than on the specific problems faced by women living in areas of industrial decline. In comparison, in the Scottish city of Dundee, there has been a similar history of women's involvement in heavy industries, in this case processing jute, a trade underpinned by Britain's colonial links to the Indian subcontinent. Here a tradition of strong, economically active women has developed which has carried through to women's participation in more recent forms of employment and, for example, women's part in the trade union campaigns over the closure of American-owned branch plants. The local culture of women's employment and political involvement was important enough to feature in locally produced theatre shows about aspects of women and work, including the following song about a woman involved in the Timex strike.

Alice's Song

I was born in 1946
Worked here for twenty seven years
Had my kids, then I came back
Toed the line and got the sack
And now I'm kept out standing on the line.

Now it's gone, I've found a different view
From my window I can see what's new
There's a rig out in the bay
Where once the ferry sailed away
And it's feeling better far from the line.

There's nothing special about this place
Another time where people lose, I guess
Just a factory on a hill
Where people found the will
To stand together on the line.

People say the city's beaten and destroyed
Say there'll be no more, we've put off those Yankee boys
Because we organised together
To make our lives get better
There'll be no more, we'll all be standing on the line.

It's just a job and you dream of something else
Some other home, some other life, some other place
You could work an age and never find
A better reason for the time
You spent outside the gates on the line.

It's just a job and you dream of something else
Some other home, some other life, some other place
You could work an age and never find
A better reason for the time
To keep on keeping on, on the line.

Ricky Ross

(The song turns on multiple meanings of 'on the line' – on the assembly line at work, on the picket line while on strike, on the line at the unemployment office.)

Involvement in labour politics may also feed into other forms of action that shape women's identities. For example, in Leipzig, a history of extensive forms of large-scale industrial production led to extreme environmental and urban decay. Women's (and men's) involvement in the state-paid employment that produced this contamination, together with their everyday experience of reproductive work in this decayed environment helped promote their subsequent political activism (as Smith discussed in Chapter 6). However, employment and labour politics are not the sole elements of significance in regionally constituted femininities. As mentioned above, an important factor influencing the construction of dominant femininities in Peru in the 1980s was the context of civil war coupled with economic crisis. At this time there was a marked difference in the ways women in rural and urban areas were affected. From the early 1980s the civil war tore through the southern highlands of Peru, taking its toll particularly on lives in Apurímac (where the case study of Andahuaylas discussed in Chapter 4 is located) and in the neighbouring province of Ayacucho (Palmer, 1992; Poole and Renique, 1992). The effects of the civil war were not felt in the capital, Lima, until bombing began in earnest in 1990. Large numbers of rural people displaced and many were forced to migrate to Lima or to provincial towns such as Andahuaylas as innocent villagers were caught between the warring actions of the army and the Shining Path (Contreras Ivarcena, 1991). The majority of those who were displaced and migrated were Quechua-speaking women who were obliged to flee their land when their partners were killed or 'disappeared'. As a result large numbers of women found themselves in female-headed households. A significant proportion of the women who enrolled in the PAIT programme in Andahuaylas were widows, some of whom had arrived in the town from remote mountain villages. In one community where fieldwork took place, all the women working in the programme had left their village together with their children under the cover of darkness after balaclava-wearing 'soldiers' had come to the village and shot all the men. The women and their children arrived in Andahuaylas (several days' walk from their village) in early 1984 with nothing. Consequently, their particular experience of the welfare later provided by employment in PAIT was different from that of many of the other women in the programme in Andahuaylas, many of whom still had a husband, and very different from that of women working in Lima at the time who had been less subject to these experiences.

Overall the case studies show that place matters in the construction and contestation of gender identities at a variety of scales from the macro-region, through the nation-state to intra-state regions. However, the different scales and the places that 'matter' are not distinct, bounded spaces. Returning to Doreen Massey's theorisation of the fundamentally interconnected nature of 'place' and Cindi Katz's and Neil Smith's questions about the constructed nature of conceptualisations of 'scale' which were discussed in Chapters 1 and 3 (Massey, 1994, 1996; Massey and Jess, 1995; Smith and Katz, 1993; N. Smith, 1993), we argue here that women negotiate their gender identities within sets of interconnecting (and sometimes contradictory) contexts and relations in different places. Processes of gender identity formation are structured through, and in turn serve to structure, powerful relations of political, social, economic and cultural control and influence that operate in these places. These processes are, at the same time, shaped in and through the relations among places. Thus the 'places' that matter are simultaneously constructed through global, national and regional relations, all of which are, at once and to varying degrees, inter-connected. So when we say that place(s) make(s) a difference, we are not propos-ing that women's identities are uniquely different within a series of bounded local places. Rather, we recognise that the constitution of both place and gen-der identities is relational, and therefore that women's experiences in/of place are constituted through a multiplicity of socio-spatial relations.

Furthermore, we want to make it clear that we are not suggesting that these processes of global, national and local change act on a unitary category of 'all women'. Even within places, but also across and between places, the experi-ences women have of these processes vary considerably, as do their responses to them. Thus, femininity is not solely shaped by one's gender and place in the world; differences among women in terms of class, ethnicity, (dis)ability, sexuality and so on are exceedingly important. In the Sheffield example, we can see that differences in class position were influential in differentiating women's experiences of the labour market and childcare system. Similarly, differences in class and ethnic background were important in Dwyer's study of young Muslim women in two schools in 'Hertfield'. In Peru there were not only differences between urban and rural areas, but also among women within these areas, for example in Andahuaylas between Spanish and Quechua speaking women. These differences among women are sometimes reflected in the physical landscape, for example in the segregation of housing by class or ethnicity, and sometimes in the social landscape, for example through the emergence of different local cultures. However, we also want to note that these differences are not only sources of distinction between women. Instead these multiple and fractured iden-tities produce a multitude of limitations and possibilities for the constitution of new femininities, particularly as women are drawn in to the wider relations discussed above in highly varied and often contradictory ways. These issues are considered further in the following section on everyday spaces.

Everyday spaces: sites and networks

The different spaces of everyday life – be it the paid work-place, the home, the pre-school playgroup, the high street, the school toilets – offer different poten-tialities for the (re)construction of gendered identities. Here we concentrate on the local, everyday spaces of paid work, home and neighbourhood, which not only include bounded, geographical sites, but also the more diffuse geographies of social networks. The constitution of each of these spaces is reinforced by social and material relations of control which work to enable and sanction some forms of behaviour and not others. As such, these everyday spaces constitute sites and networks in and through which powerful ideas about femininities operate. The ways in which women are expected to perform their femininities, and the ways in which femininities are constituted and reproduced, are dif-ferent in these different spaces. Dress, language and personal interaction, for example, which are deemed 'appropriate' in the home may be 'out of place' in the school or the work-place. Moreover, these spaces are themselves in part constituted through discourses about gender, often in relation to constructions of 'ideal' femininities associated with characteristics such as motherhood, chastity, fidelity and hard work. These discourses about women's proper role in family and society are highly spatialised, such that women are seen to have specific social roles and responsibilities in specific spaces. It is through these ideas about 'appropriate' femininities that our understandings of particular spaces become gendered.

To summarise, we argue gender and space constitute and are constituted by each other in two ways: gendered identities are constituted in and through particular sites; these sites are constituted through our ideas about gender. The dialectical relation between these two process is important not only in the reproduction of gender identities and gendered discourses about spaces, but also in changing them. The two processes reinforce each other, but equally change in one can cause changes in the other. For example, changes in the social practices shaping the use of a particular site can change the gender meanings attached to it. In other words, everyday sites and networks are also spaces where powerful ideas about femininities and the social and material relations which are associated with them can be contested and reworked. We can see these everyday spaces as 'sites' in which multiple relations of domination/resistance, meaning and belonging come together in often complex and contradictory ways. This understanding of 'sites' draws on the work of Radcliffe and Westwood (1996) who argue, based on Foucault's notion of power 'suffused' in and across a society, that no single line of conflict (such as gender, or 'race', or sexuality) wholly defines the context for identity formation. Because identities are multiple and fractured, the relations affecting particular sites are also multiple and open to challenge.

We explore these relations below through an analysis of the spaces that were particularly important in our four case studies. While our analysis here separates paid work, the home and neighbourhood, in reality they overlap. Paid work can be a space of learning, just as it can be domesticated when women take children to work; home spaces can be porous, being both potentially sites of paid work and neighbourhood action. We draw out some of these overlaps and linkages in what follows.

Paid-work spaces
The previous section on the difference that place makes described a diversity of experiences among women in relation to paid employment. Nevertheless, as Chapter 1 suggested, it is possible to identify some commonalities across the case studies in dominant understandings of masculinity and femininity which have tended to emphasise the importance of paid work for men and domesticity for women. Various dominant ideologies and practices of masculinity and femininity construct men's role as being the breadwinner and providing economically for their family and women's role in employment as secondary or supplementary to their main role as domestic carer. Work is therefore seen as an essential part of men's identity as men, such that the inability to provide economically for the family and/or unemployment are understood as emasculating experiences. In contrast, for women, many traditions of femininity

have excluded involvement with what was defined as 'work', namely paid employment outside the home. These definitions of 'work' are themselves gendered – they exclude the labour of family reproduction, which is predominantly carried out by women outside formal wage-labour relations, from considerations of 'work' while also fixing women's involvement in paid employment in relation to these unpaid, domestic forms of 'women's work'. In a variety of geographical and historical contexts, then, 'respectable' femininity, particularly for married women with children, but also for women with financial independence, has been constructed as involving an absence from formal wage-labour relations. The fact that many women, or indeed the majority of women, might not be able to achieve this ideal does not remove its importance as an ideological construct. Moreover, not only is the idea of paid work defined in many cases as masculine, the work-place itself has often been coded and preserved as a masculine site. This coding may involve the intersection of legal restrictions that enforce or facilitate the exclusion of women, of sexist organisational cultures and of informal social practices that define particular work-places as sites in which women are expected to perform particular narrowly defined femininities or to perform what is regarded as a suitably feminine version of the masculine behaviour deemed 'normal' in the work-place (McDowell, 1995). These relations are often manifest in and reinforced through the microgeographies of the work-place (Spain, 1992).

Our case studies on eastern Germany, Britain and Peru demonstrate that although the coding of employment as a male sphere is widespread across different countries, the extent of its impact and the specific forms that this takes vary considerably and are culturally specific phenomena. In all four case studies many women were in paid work, or in the case of young women were in education prior to what they assumed would be involvement in paid employment. For some women, involvement in paid employment came about as a direct response to specific economic and social crises or necessities. In Peru many women were obliged to work in order to supplement or secure a household income in a time of economic crisis. Similarly, in the Sheffield case study, some women needed to work in order to cope financially. However, for many women in both places there was also a sense in which they wanted to work for their own self-esteem. This self-esteem came in a range of forms including the knowledge that they were contributing valuable income to the household (particularly for their children), the ability among professional women to develop career paths of their own, or the knowledge that their own self-worth did not depend exclusively on their partners and children. 'Going out to work' also gave many women access to social involvement with other women or with men and a source of collective solidarity through the work-place. Undertaking paid employment outside the home was important for the construction of femininities for these women. However, it is crucial to note that involvement in paid work had already come to be defined by many women across the case studies, including some who were having difficulties in gaining access to paid employment, as a 'normal' part of their femininities. This often reflects decades of social change and women's involvement in paid employment.

The case study of work in Peru provides a clear example of how women's participation in paid employment resulted in a (selective) reconstitution of the definitions of femininities and also in turn reworked the gendering of work spaces. PAIT constituted a radically different experience for women because there was little previous history of women's involvement in the formal sector and particularly not in state employment in Peru. While some women had been involved in informal work it had been as domestics and street-sellers rather than in the construction sector, which was the work required in PAIT. Women's participation in this particular paid work offered a range of possibilities for reworking previous forms of femininity to include the identity of 'worker'. Involvement in PAIT meant women received a regular, independent source of income. It also placed women in new spaces of work outside the home, often at a distance to their own home or even outside their own neighbourhood, in which they came together to work with other women (and also with men), doing work that the state intended to be men's work and that most women understood as such. Their involvement in these public work-spaces began to redefine the gendered meaning of these spaces. However, these were partial redefinitions which still operated in relation to assumptions about the nature of 'men's work' and to the practical need for individual women to adopt particular strategies to access these work spaces that were otherwise 'dangerous' or transgressive spaces for women.

The longer-term 'normality' of work as an element in women's identities is perhaps most evident in the case of women in eastern Germany. During the communist period, the category of 'worker' was, at least officially, coded as including women and men, and indeed women constituted over half of the GDR labour force. However, this is not to say that gender did not structure the work lives of women: women consistently earned less than men before reunification since their employment from an early age was directed into particular sectors of the economy and women's access to senior posts was less than it was for men (Nickel, 1990). Nevertheless, paid employment was constructed and experienced as a normal part of women's everyday lives in the GDR. The conditions under which women were involved in this paid employment, including state patriarchal assumptions about women's domestic care, meant that many work-places in fact incorporated a range of social facilities including childcare. Economic restructuring and the shortage of public expenditure associated with post-communist 'transition' and reunification have fundamentally undermined women's access to paid work. For some women who have been able to keep or find well-paid jobs, access to a range of consumer goods and the end of the shortage economy and its demands of time and effort in acquiring everyday provisions have produced an alternative structure to the involvement in paid work, more akin to western models.

Other women find themselves coping with unemployment or radical changes to their work patterns. While some view this as relieving them from the 'burdens' of work in the GDR, most women resist the pressures on them to redefine their femininities to exclude full-time paid work (although some women are also happy to be able to engage in part-time work, which was officially discouraged in the

GDR: Rosa-Luxemburg-Verein, 1992; Kreckel, 1995). They refuse to give up their expectations that they will work, as they register as unemployed, take part in retraining programmes, change their employment, develop new 'entre-preneurial' skills, or participate in the types of workfare programme that pro-vided some staffing for the neighbourhood activism discussed in Chapter 6 (Quack and Maier, 1994). Thus in negotiating new femininities, these women draw on past traditions and resist the masculinisation of the work-force.

For many women, then, the definition of 'geographies of femininity' already includes the spaces of paid work. However, we do not want to suggest that this (re)definition of femininity is entirely voluntaristic or without cost for all women. Women in the case studies were sometimes faced with opposition from partners or family when choosing to go out to work. The young women in Dwyer's study who did undertake part-time work outside school hours often had to negotiate this through their adherence to what were considered appropriate femininities, sometimes through careful use of appropriate dress codes. The difficulties of transgressing 'traditional' gender identities and geographies are perhaps most obvious in the case of the women walking to the PAIT programme who were obliged to go in groups for fear of attack for 'being in the wrong place at the wrong time'. However, some women in both Sheffield and Peru felt dissatisfaction and sometimes guilt about not performing more domestic femininities such as caring for their families full time in the home, and some were obliged by economic circumstances to perform these new femininities when they might have preferred to perform more 'traditional' ones.

Indeed, it is important to note that the effects of any of these processes on femininities in particular sites vary for women who are situated in particular social relations of advantage or disadvantage (through, for example, class posi-tion). Holloway shows in Chapter 5 that the cultural and material resources that middle-class mothers have at their disposal, and in particular their more advantageous position in the labour market, mean that they are better able to secure entry into the work-force and maintain their position within it than working-class women whose limited access to mainly low-paid jobs means they are often unable to afford non-familial childcare provision. Differences between these two groups are also evident among those who find employment: whereas many middle-class mothers can afford to focus on family activities outside work time in order to 'make up' for not being there during the day, the common strategy among working-class mothers of looking for evening and weekend work (in order to fit in with the hours that their partners are at home and able to provide care at no cost) can reduce the amount of time families may spend together. We now turn to the home, often thought of as the dichotomous 'Other' to work environments, as a site for the production of femininities.

Home spaces
Central to many definitions of 'traditional' femininity are ideas about women's relations to others, in particular to their parents, partners and children. In child-hood girls are supposed to be 'good daughters' and in adulthood also 'good wives' and 'good mothers'. In dominant western gender ideologies the stress

on motherhood, for example, as a key element of femininity is such that women without children are either pitied for their infertility or labelled as 'selfish' or 'unnatural' if they choose to be child-free. The association of women with mothering provides a culturally specific, but strongly effective distinction between femininity and masculinity. As Holloway argued in Chapter 5, the use of the terms 'mothering' and 'fathering' in the English language distinguishes clearly between men's and women's respective inputs into children: mothering implies care and nurturing, fathering implies only a biological role in conception (although many fathers and mothers contest this in the practice of their own families).

One of the principal sites in and through which women negotiate these identities as daughters, wives and mothers is the home, a space of childrearing and daily social as well as generational reproduction (Gregson and Lowe, 1994, 1995). In turn, our ideas of home, what it signifies and how it should be organised, are (among other things) intimately gendered. The saying 'a woman's place is in the home' summarises neatly the gendered coding of home space as a site where women (are supposed to) undertake womanly activities. Dominant constructions in western society, for example, define the home as a feminine space which is a haven (for men) from work which women make, both figuratively and literally: home signifies a space for comfort and caring, and women make this through their labour. Men choosing to undertake activities in the home such as caring for children are viewed as doubly unmasculine – in relation to the action undertaken and the site in which it is performed.

The issue of mothering and its relations to home was one particular aspect of femininity that emerged from the case studies. In post-war Britain, for example, there has been a particular cultural assumption that 'good mothering' for a range of different social groups means mothering in the home. Some women in our case studies clearly believe in the importance of these feminine activities in this feminine space, and therefore reproduce these ideas through their own social practices. For example, Holloway's study of motherhood showed that some women value mothering in the home to the extent that they do not want to rework their version of mothering to include paid employment if it means using childcare outside the home. Rather, they take the decision to remain within the home and perform their identity as mother in that site, reinforcing as they do a particular constitution of what, in the British context, is seen as a more 'traditional' femininity. Similarly, Dwyer's study showed that some young Muslim women, who value the home as a site of cultural and religious integrity and reproduction, seek a role in the future as a mother who will pass on religious and cultural traditions.

This association with the home space is not, however, always something that women choose freely and it can be imposed by other people and structural constraints. The structural position of some working-class mothers in Holloway's study was such that they could not afford to go out to work because the wages they could obtain were insufficient to cover their childcare expenses. Some of these women were therefore confined to mothering in the home against their own wishes, an experience that can be extremely isolating. Association with the home is by no means restricted to the influence of relations of mothering. In a slightly different way, some of the young Muslim women in Dwyer's study were forced to maintain their association with home since they were obliged to help with domestic tasks in a way that was not true of their brothers. For such young women the home was seen as an appropriate space and their activities outside the home were closely monitored.

Similarly, for some women in eastern Germany, the radical restructuring of employment and the loss of childcare make unemployment and being 'at home' unavoidable. However, interpretations of women's reactions to these processes vary considerably. Studies of their experiences across eastern Germany and other post-communist states suggest that some women see being at home, the chance to be a housewife, as desirable because they associate it with a western style of life, and/or because being at home is a relief and a retreat from the multiple burdens that women juggled in the communist period (Nickel, 1993b; Bruno, 1995). Furthermore, popular meanings attached to the 'home' in the GDR and other communist states stressed its removal from the otherwise pervasive political influence and social controls exercised by the state (Lemke, 1991; Einhorn, 1993). Hence, these women seek to 'reclaim' a model of femininity that was not approved of in the GDR. We see from this that the meanings of 'home' are by no means fixed or universal. However, while these are important aspects in the relation between women and home, it is possible to argue the attractions of the housewife model are a *post hoc* rationalisation of the realities of post-communist transition and its selective effects on women. More common among the women in Smith's study was the type of response that argued that being at home was a restriction arising from their exclusion from the social relations and income associated with paid employment (Hildebrandt, 1996). By expanding their concerns beyond the bounds of the individual home to the public sphere of the neighbourhood, many of these women sought to resist their restriction to the spaces of the home and to constitute their femininities in relation to forms of political work that had previously been impossible, often associating their actions with extended caring relations (for the elderly or children) but also drawing on notions of women as equal political subjects. In doing so they strategically expanded the notion of 'home space' to include a concern with and a right to act in relation to the spaces of the neighbourhood and even the city.

Other women in the case studies also sought to renegotiate their relation to the home and thereby began to change its meaning and the social practices involved in its constitution. Looking first at Dwyer's study, we find examples of young Muslim women trying to reconfigure their association with the home. Some young women employed a temporal strategy – the middle-class girls in particular were keen to carry on with their education and have a career before they were in their own terms 'tied down' by marriage and family and their independent lives ended. Others took a more radical approach and used new Muslim identities to resist social control in the home, for example over dress, and to renegotiate their access to public space and wider post-school opportunities. For these women, an assertion of a self-consciously Muslim identity enabled them to take some of their rights as they defined them in their reading of Islam and resist what they saw as the 'constraints' of 'Pakistani culture'. However, their ability to do this was nevertheless hampered in some cases by the ways in which others, both Muslim and non-Muslim, read their dress and actions. As Dwyer suggests in Chapter 7, the implications of such 'new' Muslim identities for the negotiation of femininities are complex.

In Holloway's and Laurie's studies we can see that some women successfully redefined their relation to home by gaining access to paid employment (see previous section). This was achieved not by a total rejection of domestic ideologies, but by a reworking of them (for example, going out to work for one's children) which legitimated women's access to paid work outside the home, an activity that many women in both countries valued as it challenged their social isolation (see Chapters 4 and 5). Many women in PAIT, for example, spoke about the relief they felt about being able to 'get out of the house' to go to work, as they felt that home became almost like a prison. They felt isolated in their homes and in the domestic-centred social relations associated with home space which were usually limited to interactions with family and extended family members. Many women felt restricted by this situation, in some cases because abuses such as domestic violence and family power struggles could go unchecked, with the result that women's domestic isolation became a private affair. However, for some women in PAIT their experience of home-based socialisation from an early age initially made it difficult for them to interact outside the home with people who were not family members and it took many some time to become accustomed to the strangeness of being away from the domestic environment.

This reworking of the meaning of, and social relations surrounding, the home can also be seen in Holloway's study in relation to mothers' choice of non-parental childcare provision. We mentioned above the importance that is placed on maternal care in the home for very young children in British childrearing discourses. These ideas, which are part of the gendering of the home as a female space, were sometimes partially and sometimes more fundamentally reworked by the mothers in Holloway's study. For example, some mothers in paid employment chose to use childminders because they provided care that was similar to that which mothers would have provided themselves had they not been out at work. Other mothers used nurseries to emphasise that their baby-care was as much like home care as possible within an institutional environment. On the one hand these mothers freed themselves to work outside the home and negotiate a new femininity; on the other they reproduced the idea that care by women in the home is a good thing, thus reinscribing the gendered meaning of the home. Some other mothers, however, also began to challenge the idea that home care is best in a more radical way and deliberately chose nursery school rather than home-care environments for their older pre-school children. In doing so they drew on newer ideas about the importance of a pre-school education outside the home for the 'proper' development of their children. In this case then, women who were negotiating new gender identities for themselves also began to transform the meaning of home.

Femininities are, therefore, constructed, in part, through women's relations to particular sites which are often strongly gendered. However, women's identities are not only constructed in relation to what is assumed to be a feminine sphere, such as the home. They are also constructed in, through and across the distinctions among sites. One particularly important example of this is that the categories of 'mother' and 'worker' are expressed spatially in the mapping of the mother/worker dualism onto the sites of 'home' and 'work'. The dominant ideas that women's rightful place is mothering in the home, and men's rightful place is in working outside the home have had powerful consequences on, and have been reinforced through, social practices associated with these sites, which are viewed as dichotomous. Dominant constitutions of femininity and masculinity, in several of the case studies, constitute 'mother' and 'worker' as mutually exclusive identities which are stereotypically gendered to allocate women-as-mothers to 'home' spaces and men-as-workers to 'work' spaces. As we have seen, these assumptions hold in practice for only some of the women in the case studies, but this does not mean that dichotomous assumptions about the association of femininities with particular sites do not affect the possibilities that women have for constituting their femininities. Since individual women negotiate their femininities in, through and across the gendered spaces of both home and work, for example, change in one may be associated with change in the other.

Neighbourhood spaces
Neighbourhood spaces as the location for daily and generational reproduction (shops, care services, play spaces, schools) are often coded as feminine spaces through western discourses on gender, though the same is not true of all cultures. Alternative codings of such local space are most obvious in the case studies in the problems faced by young Muslim women in Britain and by women in Peru in negotiating non-familial local spaces which are defined as masculine 'public' spaces. While the notion that women in particular have a 'natural' affinity with local areas is clearly just that, notional, it is important to understand that the nature of many women's lives means that they find themselves more strongly tied to local areas than many men. Social practices surrounding the responsibility for children; cultural understandings of women as more responsible for housework and the daily reproduction of the family; greater lack of financial and other resources all mean that, for many women, life tends to be strongly related to particular local areas. Furthermore, as Smith's study suggested, the neighbourhood is also often a site of paid work for women who may choose locally based employment precisely because it allows them to balance paid work with their responsibilities in the home. We want to stress here that we are not suggesting that women do not live in wider networks. However, our case studies illustrate a number of ways in which neighbourhood space is important in the construction of 'new' femininities. One particularly important point that our analyses raise is that the neighbourhood provides a material context in which women's lives are shaped as well as a spatial context for social networks of different kinds.

A focus on the material conditions in which people live highlights the links between home and the neighbourhoods that they make up. Obviously the conditions of people's homes vary considerably between the women in our studies depending on their location in particular neighbourhoods, towns and their positions in the world economy. They range from self-built homes in Peruvian shanty towns to well-appointed, comfortable housing in middle-class neighbourhoods in the British case studies. The 'home' therefore constitutes a (more or less temporary) anchoring of people in particular places and social relations, often most immediately in particular neighbourhoods, where social relations of advantage and disadvantage are reinforced or disrupted through access, for example, to particular facilities, such as schooling, housing or employment. The physical nature of the home may be a basis for women's wider neighbourhood actions. For example, for many women in the Leipzig case study their homes' physical condition (or lack of condition), the property relations governing them and the emotional attachment and effort that had been invested in their homes were significant elements affecting the actions they were moved to take, both individually and collectively, in the neighbourhood. This example underlines the point that the 'home' is not adequately defined by its dichotomous relation to work. It is rather, like the neighbourhood, a site that is multiply constituted, sometimes quite literally, by a series of relations. Activism by the women in Leipzig around housing conditions and local environmental problems involved challenging the parameters of these relations, pointing out the limitations of the abstract notions of tenancy rights or of physical planning, which often operate on the bases of universal and abstract subjects. The women regularly criticised pronouncements on tenancy rights for being divorced from the messy complexity and great uncertainties that women faced in their homes, and they challenged local development plans which treated homes and their tenants as largely interchangeable and ignored the social context of residency in particular neighbourhoods.

As important as the physical conditions of neighbourhoods is the construction of neighbourhoods through local social networks or communities. These networks, like the physical, mappable resources of an area, may constitute a set of potential resources, connections or constraints through which new femininities are shaped. Dwyer shows, for example, how young Muslim women's use of public space is monitored and controlled through local social networks. In this case it is not only the family but the local Pakistani community that measures a young woman's respectability by the manner in which she uses public space, filtering information back to parents or members of the extended family, and forming the young woman's reputation within the wider community. As Dwyer illustrates, young women respond to these discourses and surveillances in a variety of ways, sometimes conforming to neighbourhood strictures and sometimes acting in ways that subvert them. It is, of course, not only Muslim girls who are subject to moral discourses that link behaviour in public space to reputation – as Dwyer emphasises in Chapter 7. The importance of local moral communities is also evident in Holloway's work on childcare cultures in Sheffield. Here, the two case-study areas of Hallam and Southey Green were characterised by different moral geographies of mothering, that is different sets of ideas and practices about what makes a good mother and what constitutes good mothering. These ideas are (re)produced in part through local social non-familial networks, particularly in the middle-class area where these networks also acted as material sources of help, for example in terms of baby-sitting, as well as advice (Holloway, 1998c).

Networks of local action may also be created in ways that set out explicitly to develop possibilities for change among women. The coding of the local as feminine has led some writers in geography and elsewhere to suggest that women are often more likely to be active in what might be called community politics, and also in more formal political involvement in their local area rather than at wider scales (Brownill and Halford, 1990; Jelin, 1990; Radcliffe and Westwood, 1993; Afshar, 1996; Graham and Regulska, 1997; Stephen, 1997; Regulska, 1998). Some of the women in our case studies are indeed involved in community politics. These politics may sometimes reinforce the gendering of neighbourhoods as a feminine space and may sometimes draw on alternative identifications. In the Sheffield case study, the activities of some middle-class women in producing a childcare magazine called *Chatterbox* and in running local playgroups were a form of community politics that both empowered those women and shaped the lives of other mothers in their neighbourhood, helping them carry out mothering work more effectively and reproducing particular ideas about 'good mothering'. This activism was based on the understanding and social reality of women's responsibility for childcare and reinforced both this and the idea that the neighbourhood is and should be a woman-friendly family space. In contrast women's experience in the PAIT programme made local spaces women's spaces in a way they had not been before. Women were made physically present in local space, though they also had to adopt particular strategies such as travelling to and from work with other women to negotiate their passage through public space. The schemes also developed women's networks with other women, allowing them to act in communal ways.

Far from assuming that local activism is only related to 'gender issues', the case studies suggest that aspects of women's identities other than gender *per se* may be reinforced or transformed through local activism. In Smith's study of eastern Germany, for example, the fact that women wanted their community activism in Leipzig recognised reinforced a sense of their identities as eastern Germans. Even though, in other contexts, these women were quick to reject the need to choose between an 'eastern' or 'western' identity, their appeal for political recognition was based in part on their particular experience as eastern Germans and as women. However, their account of this identity was not an exclusive one and they also drew on experiences of housing-based action, for example, which are common across the boundary between 'East' and 'West'. Thus neighbourhood activism by these women at once reinforces a regional rather than a gendered identity and simultaneously works to restructure what that regional identity might mean. In Lima, Peru, on the other hand, it was only in the context of meeting women from adjacent communities that women from José Galvez (one community in the south of the city) got to know each other for the first time. So while working in PAIT strengthened common bonds between women it also reinforced identification with specific bounded communities. This dual strategy of negotiating identity in place and in relation to wider networks was also found among the young women in Dwyer's study. Many of the young women interviewed were consciously doing the 'work' of negotiating multiple aspects of identity, for example, by working together on the fashion show or participating in the actual discussions carried out in the research (see Chapter 3) to explore their own identities and to communicate them with their classmates. In doing so they sometimes selectively drew upon, renegotiated or rejected different elements of diasporic identities, stressing, for example, Pakistani, British or Muslim identities while negotiating them in the specific sites of everyday contact.

Finally, we would argue that many of the 'spaces to learn' that the women in our case studies have used and even created are often (though not exclusively) found in local, neighbourhood spaces. The example from Holloway, above, illustrates how neighbourhoods can provide spaces to learn, in that case learning about mothering through mothering networks. The women who were involved in the PAIT workfare scheme in Peru were also able to find spaces to learn. They learned technical skills such as paving from the men in the programme and were able to use these to improve their own homes. In Andahuaylas several women joined together to paint and plaster each other's houses and one woman even spoke about how surprised her husband was to come home and find the house refurbished in his absence. Neighbourhood activist groups in Leipzig provided spaces and events at which the women could learn about the new legal and political frameworks that had been introduced, such as tenancy and property rights or the regulation of different forms of democratic participation and citizenship rights. They also saw it as their role to teach other people about the possibilities and limitations of these new circumstances, and to explore collective strategies for contesting the processes of urban change. The schools in Dwyer's study provide the most formalised spaces to learn, though what the young women learned about their gendered identities came as much through the informal curriculum of the school as it did through timetabled lessons. The space of the school was contradictory, sometimes providing a space where other people's stereotypes had to be resisted, sometimes a space for new religious identities and sometimes a space for experimentation and imagination – a space to dream. The complexity of this site illustrates the shortcomings of cultural conflict models that suggest that the home is the space of cultural integrity, while the school is merely a secular environment.

Overall, the case studies suggest that any fixed notions about the gendering of particular spaces are likely to be overly simplistic. The everyday sites and networks of paid work, home and neighbourhood are powerfully gendered in different ways in different times and places, both as ideas and as they are constituted through material and social practices. The relations that shape these sites vary among different places and are also open to change through the workings of large-scale processes of economic, political or cultural change. We would argue that women, in forging 'new' femininities, are both shaped by these geographies and reshape them, changing social practices and the meanings of these spaces. The constitution of particular sites is not fixed. Indeed the very fact that sites are constituted in relation to other sites (home in relation to work, etc.) means that change in one can produce multiple and/or contradictory change in others. As the practices and meanings associated with particular sites change, new possibilities, opportunities and problems emerge for women in shaping their femininities. In turn different women may, in reworking their femininities, rework the constitution of these everyday sites in a variety of ways. The implications of these processes are considered further in the following section in which we discuss the gendering of geographical change and geographies of gender change.

Reconfiguring change

So far in this chapter we have shown that the three-fold typology of the ways in which geography is significant in the construction of femininities (set out originally in Chapter 1) – which highlights the difference that place makes, the importance of everyday spaces, sites and networks, and the significance of the ways in which these spaces, sites and networks are themselves gendered – provides a productive starting point for our analysis of the geographies of new femininities. In working through the implications of our case studies above, we were able to develop this typology further by incorporating questions of social change and the multiple constitution of identities. In doing so, our analysis brought out the importance of the relational constitution of global/local, the dialectical relation between spaces and their gendered meanings, and the often problematic but nevertheless fundamentally important actions of different women in different types of spaces as significant elements in the geographical constitution of femininities.

In the remainder of the chapter we now deal explicitly with the question of how we understand change. Two particular themes are used to explore this somewhat abstract idea. First we consider how the analysis of the case studies leads us to gender the geographies of change. In doing so we point to limitations in the ideas of space and place implicated in the discourses and practices of economic, cultural and political change and we suggest some distinctive understandings of the geographies of change that emerge from these studies. Second, in the section entitled 'Placing changing femininities', we work through the understandings of space and its relation to identity which are implicated in the range of ways that identities can be understood as changing. In particular we discuss how the need to deal with multiple identities is problematic but also offers possibilities for negotiating new femininities. This then leads on to a final discussion in Chapter 9 of how we might understand the femininities described in our various case studies as 'new' femininities.

Gendering the geographies of change

A gendered perspective on geographical processes of change highlights the limitations of the understandings of space and place employed in several key discourses and practices and also signals some possible ways forward which integrate gendered and geographical perspectives on change. We begin here by exploring how a focus on changing gender identities complicates and disrupts the 'smoothed out' spaces of globalisation, neo-liberalism, transition and similar discourses. We argued in Chapter 2 that hegemonic discourses about major processes of change (such as globalisation, development, post-communist transition or neo-liberalism) share a tendency to universalise their effects. This makes changes that occur in relation to them appear both inevitable and (by and large) desirable (since they produce a better market system, or constitute 'progress', for example). One of the ways in which this assumption operates is through assertions that space is 'overcome' by these powerful changes. In their hegemonic and universalist forms, these discourses claim to operate in ways that are somehow outside spatial relations and that simply inscribe their powerful outcomes on passive spaces, erasing the distinctiveness of places. Gibson-Graham (1996) illustrate how the conceptual frameworks used by geographers and others in relation to capitalism, for example, reinforce these understandings by constructing capitalism (and therefore the social and economic changes associated with capitalist development such as globalisation) as 'all-powerful':

> It is through these discursive figurings and alignments that capital is constituted as large, powerful, persistent, active, expansive, progressive, dynamic, transformative; embracing, penetrating, disciplining, colonizing, constraining; systemic, self-reproducing, rational, lawful, self-rectifying; organized and organizing, centered and centering; originating, creative, protean; victorious and ascendant; self-identical, self-expressive, full, definite, real, positive, and capable of conferring identity and meaning. (Gibson-Graham, 1996: 4)

A variety of authors, drawing to varying degrees on post-structuralist arguments, claim that the ways in which these and other macro-discourses of change have been imagined make it difficult to conceptualise alternatives to their universalistic and often strongly normative claims. Alongside Gibson-Graham's feminist analysis of capitalism, other writers have focused on discursive accounts of processes such as 'globalisation' (Morely and Robins, 1995), 'development' (Escobar, 1994; Crush, 1995) or 'post-communist transition' (Smith and Pickles, 1998). Their analyses, although varying considerably, share a concern with the ways in which the very discourses of these processes tend to overstate their cases, drawing all relations into their spheres of influence. These discourses are shown to be powerful forces in shaping not only theoretical understanding but also the material and social outcomes of policy and practice. One example of relevance to the case studies here is the discussion of how the two key discourses of 'development' and 'modernisation' have influenced much of post-war planning and economic policy towards 'developing countries' (note the way countries become labelled by the very terms of the discourse). Escobar (1994) has shown how the post-war boom in 'development planning' was as much about establishing a specific discourse of 'progress' towards modernisation as it was about promoting sets of practices and policies. The political message of Escobar's analysis is that there is a need to try to imagine an era of 'post development' where development goals do not frame the ways in which all other understandings of society and economy are viewed, so that measures of what is 'useful' or 'desired' are not only framed in terms of development and 'progress'. This is particularly important when it is understood that the definition of 'progress' or 'development' is dominated by western, neo-colonial agendas. In relation to gender, in particular, we can see the importance of Escobar's argument when we try to unpack the extent to which gender issues are only designated as important when they contribute to these macro-level agendas. These same development and modernisation ideas also influence the 'neo-evangelistic rhetoric' (Derrida, 1994) of those discourses of post-communist restructuring that ignore the possibility of different trajectories of change at national and regional scales. Such discourses also fail to address the massive social costs of such change which are, for example, disproportionately carried by women, as well as the possibilities that either different local histories or currently diverging politics of transition in fact produce divergent outcomes which are not merely 'failures' of transition, but which represent valid alternative forms of change (Bridger *et al.*, 1996; Smith and Swain, 1998).

A specific focus on the changing constitution of gender identities disrupts these all-too-smooth interpretations of the spatialities of globalisation. We want to reinforce here our earlier extended discussion of the importance of place in the gendering of the processes of globalisation. The corollary to the claims of hegemonic discourses of change to universality and ultimate 'triumph' is that the distinctiveness of local places is seen as having significance only in relation to predetermined outcomes of change. This is true whether the outcome of such change is welcomed or critiqued. Many political economy understandings of the effects of globalisation, for example, discuss both the shift of power to the global scale (by capital in particular) and the increased importance of the local as they 'hollow out' the power of the state (Jessop, 1993). However, the local in this case is generally seen as a focus for attempts to withstand globalising processes or for attempts to exploit difference in order to pin down capital in particular places. Both responses are seen by some commentators as futile because they both buy in to the very logic of globalisation (Peck and Tickell, 1994). Yet, as we have seen, the notion that the local 'reacts' to universal and powerful forces, while true to some extent in many cases, does not explain all forms of difference, nor are all forms of local change subsumed in these meta-processes of change.

In relation to changing femininities in particular, we obviously see it as important that the 'impacts' of major changes on women are considered, and any such consideration is an improvement in some areas of policy formation which otherwise ignore the existence of difference and the problem that women particularly might be adversely affected by change. However, there are also dangers in analyses of the 'effects' of major changes, such as those brought about by the adoption of neo-liberal policies or the operation of other political, cultural or economic aspects of globalisation, which remain 'impact oriented', that is, they concentrate on identifying the 'effects' of change 'on' women. One danger is that the size and nature of categories of 'vulnerable' women (such as female-headed households) are overstated to the detriment of other marginalised groups (Varley, 1996). This focus on restructuring and households has been critiqued for reinforcing the idea of 'capitalism' as monolithic (see Gibson-Graham, 1996). In addition, there is a need to be aware that gender analyses of change often construct women in restrictive ways as they associate them with stereotypically gendered spaces, such as the household. We hope that the first section of this chapter adequately illustrated the multiplicity and complexity of change as well as the active roles that many women adopt in contesting the formation of new gendered identities. However, as we mentioned in Chapter 2, it is also important not merely to replace conceptualisations of women as passive 'victims' with one of women as 'resisters'. Seeing women only as resisters, fighting against global processes in their local spaces, implies that women automatically embody progressive attitudes and assumes that women are seldom complicit with agendas such as those proposed by neo-liberal policies (Laurie and Smith, 1998). Such interpretations tend to homogenise 'women' as a group, and leave little scope for the fact that the 'impact' on a variety of women can vary substantially. They also overstate the power of economic and political processes acting 'from the top down'. They ignore the kinds of intersection with cultural processes, with local histories and with individual and collective actions that we have highlighted, and sideline the importance of the new spaces being opened up by widespread changes where gender identities are being renegotiated. They shift attention away from the complex ways in which formations of gender identities in local contexts interact with a range of globalising processes, and the ways in which gendered identities are formed and reformulated through a dialectical relation to the changing material constitution of everyday spaces and their gendered meanings.

We now turn to consider a second aspect of importance in rethinking the geographies of change – namely the relation between hegemonic discourses and macro-structural divisions of space. There is an apparent contradiction here between the way hegemonic discourses on the one hand claim to operate across space, even to overcome space and the specificities of place, and on the other hand divide the world up into what appear to be quite distinctive spatial units. Familiar macro-structural divisions are those such as First World/Second World/ Third World, North/South or West/East. These divisions are, however, largely viewed as fixed and generalised spatial categories through which existing and restructuring inequalities can be understood and explained. Crucially, they allow the discourses to account for difference at a general level while still maintaining their claims to be universal and/or inevitable processes. Effectively macro-structural divisions become defined as spatial expressions of the different stages of a process that will produce similar outcomes in the long run. The ultimate outcome of these processes is therefore fixed and the assumption is that each region can achieve, through the 'appropriate' adjustments, the forms of development achieved by First World/Northern/Western regions. Any failure to achieve these forms of development in practice is put down to the inability of different countries or regions to adapt in the 'correct' way.

Here, however, we want to draw out the point that a focus on the multiple ways in which femininities are constituted challenges the temptation there is to leave these universalistic claims and their fixed divisions of spatial and social processes unquestioned. This is not to deny the importance of large-scale processes of change for the formation of gender identities, or to dispute the fact that these processes are often structured through large-scale divisions in relation to the categories mentioned above. The first half of the chapter clearly illustrated their significance in the geographies of new femininities. However, here we specifically want to disentangle how these ways of thinking about change and geography are also challenged by a focus on gender identities.

Taking as one example the need to think again about the categorisation of the world into First World, Second World and Third World, many of the similarities in processes of globalisation and neo-liberalism affecting the construction of femininity challenge us to unpack the significance of these divisions. At an empirical level the case studies suggest that there are important commonalities between the ways in which so-called 'poor' women in 'First', 'Second' and 'Third' World countries construct femininities. For example, in Lima, Andahuaylas and Southey Green, money earned from women's wages makes an important contribution to the budgets of all 'poor' households. In these contexts poor women in both Peru and the UK have redefined their femininities and their responsibility as women to include providing economically for their children. In the Peru case study detailed examples are given about how this argument is presented to male partners in order to gain support for female participation in what was originally designed as a 'male' work programme. In eastern Germany similar workfare-type initiatives have also experienced high levels of female participation although without the associated opposition from male partners because of the long history of women's participation in public sector and state-led employment. The survival strategies (and therefore the 'activism' of these women) also has similarities, being based in soup kitchens, local environmental action and playgroups, all of which, it could be argued, are part of what we defined in Chapter 2 as new social movements. Therefore, the case studies show that a reworking of 'traditional' visions of femininity, and in particular maternal responsibility has been common to responses to economic restructuring across the 'First', 'Second' and 'Third' worlds, but that the precise details of these processes vary considerably. Thus it is clear that the distinctions and hierarchies suggested between these macro-regions are disrupted and cross-cut by a consideration of gendered identities. They clearly have some impact on the different life chances of women in these different places, but a detailed engagement with the precise ways in which they are implicated in constituting femininities provides a useful corrective to the tendencies of hegemonic theories to universalise and postulate normative and inevitable outcomes to change.

As a second example, the notion of a discrete 'East' and 'West' was a fundamental element in Cold War politics (Borneman, 1992; G. Smith, 1993). Now the cultures and spaces of 'East' and 'West' are still powerfully shaping the way in which social and economic change is envisaged in post-unification

Germany (Hörschelmann, 1997) and in EEFSU in general. In fact women also use these particular geographical understandings to work against or to reshape in specific ways the processes of change. They contest the definitions of 'the East' by asserting their common links to other people and places in the 'West', working across the East/West divide, and they also reassert a more 'bounded' notion of the 'East' as a political strategy for asserting the equal value but distinctive forms of their experiences as women who have experienced communism, 'revolution' and western liberal democracy. Geography is therefore not simply an outcome or a stage on which change works, it can also be a tool that is used strategically by people, often in sophisticated ways which many theoretical stances fail to capture.

In addition to our argument that a focus on gender identities requires us to rethink dominant discourses and categorisations of change, the case studies suggest not the 'withering' of the state, but the continued, though often rapidly changing, importance of the power of both the national and the local state in shaping possibilities and limitations for the reworking of femininities. One obvious example is the way in which welfare reform is affecting countries in the North and the South, with strikingly similar programmes being implemented across a range of countries. 'Workfare' initiatives, that is coupling state welfare payments to work programmes, are found in each of our three case-study countries. The programme in the Peruvian case study is not dissimilar to the programmes recently instituted by the Labour Government in the UK and has similarities to programmes in the USA or those in eastern Germany. All have attracted large numbers of women. The popularity of such programmes indicates that even though the nature of welfare is changing in the context of the roll-back of the state and changing economic fortunes, the state is still seen as the agency directing structural adjustment packages (even if it is at the behest of international agencies).

It is also the state that defines who are the vulnerable in need of protection at such times and shapes the provision of services at the local level, though this is often mediated through the politics of the local state. In the UK as a result of these shifts we have seen the definition of single mothers, for example, shift dramatically in recent years. McRobbie (1994) argues that the Conservative Party attempted to set off a 'moral panic' about welfare dependence in the summer of 1993 in the context of rising numbers of single mothers who were increasingly less likely to be employed and hence more reliant on state benefits. The trend towards more, rather than less, dependence on the state was in opposition to the neo-liberal bent of conservative economic policy, which focused on the control of public expenditure (Millar, 1994), and the conservatives' linked ideological concern to break the post-war welfare state consensus (McRobbie, 1994). A 'moral panic' about welfare dependency would thus legitimate a dismantling of aspects of the welfare state as well as government inaction in other areas. As Warner (1994) puts it:

> chronic scaremongering about female behaviour – about wild sexuality
> and aberrant maternity – distorts understanding and sinks matters of urgent

social policy – the proper provision of child care, tax reform, job training and retraining, nursery schools, housing, play areas – into a quagmire of prejudice. (Warner, 1994: 13)

Under the Labour Government elected in 1997, rhetoric surrounding lone parents (note the change in terminology) has changed: rather than being portrayed as women of 'loose morals' who 'scrounge' from the state, lone mothers are being represented as women struggling to provide for their children economically and to give them the right start in life in order that they become integrated members of society. Linked to this change in rhetoric has been the presumption that 'good' lone mothers will want to work, a presumption that legitimates attacks on the welfare system. For the first time in post-war British social policy, 'good mothering' and paid employment have been ideologically linked, though as Chapter 5 suggests childcare provision has a long way to go to catch up with these changing discourses. In comparison, in a reversal of the presumed greater responsiveness of western states to gender issues, the influence of 'Gender and Development' approaches in policy in parts of the so-called Third World, which seek to 'add gender in' to the development process, has meant that such women have long since been referred to as 'female heads of household' or 'women maintaining households' rather than as 'single mothers'. Development policies are therefore in many cases geared towards a recognition of mothering and economic provision as elements of the femininities of most women (Laurie and Smith, 1998).

The formulation and implementation of state policy can often universalise processes, using generalised, mutually exclusive categories of people and places in ways similar to the processes we identified earlier in relation to discourses of global change. James Scott's (1998) recent analysis of the 'failures' of development programmes shows in detail how processes of naming, mapping and labelling, carried out by states in legislation, policy and institutional practice, serve to homogenise diverse social systems. They seek to use broad categorisations of social groups (such as female-headed households, or lone parents) and of social processes (such as unemployment or crime) to formulate and implement policy. Scott argues that state-led discourses have had a similar effect at the national scale as the global language of neo-liberal change, for example, in marking change at an international scale. Both undervalue local and diverse knowledges and experiences. They also ensure that 'the local' or 'the indigenous' are nearly always defined in relation to discourses of homogenisation and standardisation. In other words they require a diversity of individuals and groups to position themselves in relation to these broad categorisations.

Perhaps one of the strongest homogenising tendencies relating to the state is found in questions of national identity. However, Dwyer's study suggests one example of how we might think differently about national space, challenging fixed and bounded ideas of space in the notion of diaspora. Diaspora space undermines and reconfigures an understanding of bounded national spaces through its emphasis on transnational linkages, connections and belongings.

Diaspora populations are those that result from processes of both voluntary and forced migrations and diaspora peoples are dispersed across geographical boundaries and may call several different places 'home'. Rather than working with dichotomous notions of 'home' and 'foreign' space, for example, diasporas are characterised by the transformation and reworking of cultural identities through processes of syncretism and fusion (Hall, 1992a). Dwyer suggests a range of ways in which gender identities in diaspora populations may disrupt notions of a singular or homogeneous sense of belonging. On the one hand 'new' identities might be created that are a fusion of different cultural identities – the creation of complex, hybrid British–Asian feminine identities is one example. Chapter 7 showed how many young Muslim women in Britain seek to transform the cultures to which they 'belong' rather than let themselves be 'caught between' them. However, the dimensions of identity are far from exhausted by the binary of British and Pakistani in this case as is seen where young British Muslim women seek an alternative religious identity from their parents and find inspiration in the wider Muslim diaspora, looking to Saudi Arabia or Iran, for example, rather than to Pakistan for feminine role models. Thus the theorisation of gender identities in relation to diaspora provides another means through which our studies of femininities disrupt dichotomous and bounded notions of space.

The case studies illustrate a variety of more complex relations between women and the state which reflect the complexity of the relations that constitute gender identities. Where the hardships of neo-liberalism and economic crisis combine with patriarchal gender regimes, it is unsurprising that the state overlooks women's needs and that women are often the 'victims' of change. However, women may be complicit with projects of economic reform, even where they are victims of it, while at the same time they resist other projects of the state, such as those around nation-building. For example, women in eastern Germany may favour being part of the reunited German state even though they are also in some senses 'losers' of reunification, while they also resist discourses and practices of reunification which dismiss all aspects of the former communist state. Similarly, women in Peru support the idea of state-led employment but are critical of the corruption of the specific regime that carried it out. As members of diasporic community, the young women who are the subject of Dwyer's study are skilled negotiators of the 'in-between' spaces of British–Pakistani–Muslim identities and they resist the prescribed dichotomous alternatives of belonging to one fixed nation state (see Dwyer, 1991). A focus on women's complex relationships with different state projects therefore illustrates how women's constructions of femininities work to undermine what start out as homogenising and standardising national projects. Processes are not only defined from the top down. Local actions may also (re)shape national or even international processes. While the case studies support Scott's (1998) focus on the power of the state, they also suggest that questions of whether the power of the state is in decline are not as relevant to the study of emerging femininities as the need to distinguish the particular, and changing, roles and power of the state at different scales.

When processes of macro-social change are reconsidered in relation to the multiple constitution of (new) femininities, it becomes obvious that the 'smoothed out' notions of space and the fixed categorisations of identity and space through which they script and understand the world have their limitations. By omitting both the possibilities and problems of difference, of the interconnected nature of shifting processes of identity formation and the constitution of space(s), dominant discourses of change such as those of 'globalisation', 'development' or 'transition' reflect narrow notions of the significance of gender and space and only promote 'fixed' notions of change. The following final subsection of this chapter discusses the flip-side of these issues, which is that different interpretations of how gender identities change also operate with specific 'imaginaries' or conceptualisations of the category 'gender'; of its relation to other identities; and importantly for us here, of the connections between 'gender' and 'space'.

Placing changing femininities

We have argued schematically in the introduction to the book that particular spaces are gendered when dichotomous understandings of masculinity and femininity are mapped onto supposedly mutually exclusive sites, such that paid work space becomes a man's world, and home and (in some cultures) neighbourhood become women's realms. These dichotomous understandings of masculinity and femininity, and of male and female spaces, are also evident in popular understandings of changing gender identities.

Debates such as those about new femininities, which we introduced in Chapter 1, often frame change in masculinity and femininity either in terms of a rejection of prescribed gender identities, and the spaces associated with them, or in terms of hyper-masculine or hyper-feminine performance of 'appropriate' gender identities. Thus women's liberation (whether this is viewed in a positive or negative light) is sometimes seen to occur when women claim their space in a man's world, for example by undertaking full-time employment in a formerly male-dominated occupation, becoming in the process 'one of the boys', or honorary men (McDowell, 1995). Alternative interpretations of liberation which reject these 'male' values and instead revalue and celebrate attributes associated with femininity and female spaces, which can be seen in examples as diverse as lesbian chic or the women's peace movement, depend on equally fixed, binary understandings of gendered identities and gendered spaces. The case studies discussed in this chapter include several instances where women transgress the boundaries between 'men's' and 'women's' spaces, for example rendering women 'out of place' in the 'male' spheres of paid work; equally, there are examples where women use or 'reclaim' spaces seen as 'women's spaces' such as playgroups, girls' toilets and community activities as significant sites for their own actions.

However, these 'fixed' understandings of space provide only a limited interpretation of the connections between identity and spatiality. The three-fold typology of the connections between geographies and femininities, first presented in Chapter 1 and developed further through the course of this chapter in relation to our four case studies, draws our attention to a more diverse range of possibilities as women negotiate the everyday expressions of social change in a range of spaces and places. Of particular relevance here is the fact that the social and material relations that shape gender identities in particular spaces, and our gendered understanding of those spaces, are not forever fixed, but are open to change as women renegotiate the meanings of paid work, home and neighbourhood in a range of places. This point is important as it highlights the limitations of some political and academic analyses which prioritise either reversals or revaluations of polarised gender identities and spaces but which pay consequently less attention to reworkings that simultaneously work within and between them. This point is perhaps best illustrated by example.

Laurie (1999) discusses how some Peruvian feminists as well as some of those involved in NGOs criticised the work of PAIT and only regarded women as transgressing the bounds of existing femininities when their enrolment in 'popular feminisms', based around women's domestic roles, made private (unpaid) activities public. For such critics (see Paredes and Tello, 1988), the mass employment of women in PAIT did not reflect women's transgression of traditional gender identities (which is one of the interpretations that Chapter 4 suggests). Rather it involved the co-option by the state of women's grassroots mobilisation, which had already successfully revalued the significance of women's domestically based actions in the soup kitchens or glass of milk a day schemes, into existing patriarchal structures of state action and formal employment relations. They were therefore highly sceptical of the programme's role in changing women's lives. Laurie (1999) argues that such interpretations use polarised notions of gender identities and space. A range of writers have called for economic development and capitalism to be re-envisaged so that the strict divide which sidelines 'informal' activities as both non-masculine and non-capitalist is overcome (Redclift and Mingione, 1985; Gibson-Graham, 1996). Laurie suggests that it is somewhat ironic, then, in the context of a so-called Third World country such as Peru, that the success of popular feminisms in revaluing the realm of women's 'informal' activities has been accompanied by the demonisation of 'formal' state-led employment by its critics, whereby its potential as a site for newly emerging femininities is dismissed. In doing so, Laurie argues, women's strategic choices about how they spend their time become sidelined along with the esteem they gain from paid work, the new networks they form and the connections they forge between paid and unpaid activities, all of which were discussed in Chapter 4. Her analysis in that chapter suggests that these 'fixed' notions of what might constitute change, both in relation to gender identities and the spaces in which they are performed and negotiated, are overly rigid. More fluid conceptualisations of how change might occur allow a range of possibilities to be considered, where femininities may be configured in, through, across or between categories of gender and space, often in complex patterns of stereotype reproduction and subversion.

The second point of note in this subsection is that, as our analysis of the everyday spaces presented earlier in the chapter illustrated, the spaces of importance to us – the home, paid work and neighbourhood – are not defined exclusively by gender relations, but by the intersection of these with other social relations. The 'home' for example is a site influenced by a multiplicity of factors including gender but also covering ethnicity, sexuality, national identity, religion, class and so on. Recognition of this brings other questions to the fore, most notably about the positive potential or, alternatively, the limitation that the fracturing of identities provides in the constitution of new femininities. We have already outlined some of these key theoretical issues in Chapter 2, where we illustrated in particular how early debates about the fracturing of identity became concerned with hierarchies of oppression and tended to produce list-like sets of identity attributes, but have more recently focused on the subjectivities occupied by individuals or groups of gendered, racialised and classed (and so on) people at particular moments in particular places.

As we have seen, it is clear that gender cannot adequately be understood in isolation from other aspects of identity and the ways these axes of difference combine in specific sites, spaces and contexts. These may well disrupt the extent to which women share common experiences around the identity of gender. For example, Dwyer's study emphasises the differences among young Muslim women in terms of both class and ethnic background as well as in adherence to a religious identification. There are situations where forms of social identification other than gender – for example as Muslim in Dwyer's study, or as 'eastern' German in Smith's study – cross-cut gendered forms of identity and belonging. While in some cases these 'other' forms of belonging may undermine the ability or willingness of women to choose and reshape their femininities, in other situations, women may be able to utilise the intersection of gender with other aspects of identity to negotiate their multiple and even fractured identities. Returning to Dwyer's example, for some young women emphasising a self-consciously religious identity was a strategy to combat gender restrictions which they saw as coming from being part of Pakistani culture. For others, their identification as an Asian or British young woman was more important than their identification as a Muslim and some consciously resisted the femininity and religious identity they saw as being attached to a 'new' Muslim religious identity.

The identities on which women draw are therefore not always open and fluid. In fact women may draw distinct boundaries between different groups of women themselves. Holloway noted that one of the things that divided women in the Sheffield case study, even from other women who lived locally and who came from a similar background, was the way in which they used judgements about 'appropriate' femininities to define groups of women: as one of the interviewees said, 'there is a them and there is an us'. Similarly, in the case study from Peru, women identified differences among themselves on the basis of being 'good' or 'bad' women (this is discussed further in Chapter 9). This sometimes reflected strategic representations to husbands or others and at other times merely reproduced prejudice. The boundaries women draw among themselves reflect various types of differences, some of which are based on racism and anti-communism as well as on notions about appropriate sexual behaviour, whether it be homophobia in the PAIT work-place or the subtle policing of peer group pressure which defines the boundaries of moral and immoral behaviour. What is important to note here is that the identities that are significant for women do not necessarily operate through sets of fixed social categorisations, although women do also draw on such categorisations in defining their and others' identities.

Rather than viewing 'gender' or 'class' or 'ethnicity' as fixed social categories, then, it is possible, following Hall (1992b) and others (see Chapter 2), to argue that identities are better understood as being in process, constantly being made and remade in different times and places. In this way it is no longer always important to distinguish women's struggles over power as based on *either* gender *or* class *or* ethnicity *or* sexuality (and so on) because people's identities may be multiple and shifting and, importantly, the struggles they face may be 'interpersonal and may not necessarily involve affiliations with a group' (Gibson-Graham, 1996: 19). For example, in Holloway's study of mothers in Sheffield, most women did not self-consciously define themselves in terms of class positions or class relations (for example, distinguishing what they might do as a mother from what a mother of another class might do) but rather in relation to their own local networks, moral communities and encounters with family and peers (this is not to say that their class position was not an important influence on their lives). In Smith's study, too, the women involved in urban activism generally defined their own actions in relation to the problems of urban change and the impacts it had on their work and neighbourhoods rather than in terms of pre-given social groups or categories. At the same time their actions also sought to redefine several key definitions around political involvement that were of significance, particularly that of 'citizen'.

It is important therefore to conceptualise change by understanding how 'either/or' categorisations of gender identity and of space shape processes of change and, crucially, how by moving beyond these categorisations to assess their limitations we can offer alternative and, we would add, richer interpretations of how gender identities change and how these can be understood in relation to their geographies. Dwyer's examination of gender and ethnicity attempts to do this by critiquing the cultural difference model that defines 'British Asian' culture in relation to hegemonic understandings of 'British culture' and emphasises that British Asian young women are working within and between these dichotomous understandings of two discrete and oppositional cultures. By focusing on the fluidity of identities and the complex flows between home/school, west/east, Pakistani/Muslim/British, Dwyer gives primacy to hybrid identities and focuses on the 'spaces of in-betweenness' (Kobayashi, 1994) as the main site of identity formation for young Muslim women on the verge of adulthood. The young women also negotiate these 'in-between' spaces themselves whether by challenging or subverting dominant gendered expectations at home or school or by carving out alternative, real and imagined spaces (like the school fashion show stage or the girls' toilets) within which new identities can be explored. Thus, while some intersections of different identities work to constrain choices and opportunities for women, other intersections open up precisely those new spaces and opportunities for identity formation that offer the possibility for new constitutions of femininities. The negotiation of changing femininities therefore occurs across, on and between the boundaries of different and shifting categorisations of identity and space.

The differences among women, and some of the contradictory ways in which femininities are made and remade, also raise important questions for feminist politics. Most fundamentally they raise the question of how alliances are and can be made to effect political change. As Holloway (1998a) argues, there is often a tension between strategies for political action that recognise and work with differences and those that advocate political action on the basis of shared social attributes – such as gender or class. The latter position often seeks to subsume differences to achieve political ends. Our case studies illustrate that individual and collective change in femininities operates through multiple spaces and shifting alliances. Yet these alliances are not simply chosen 'at will' because, as our case studies have shown, space matters: the constraints and opportunities to develop different gendered identities vary within particular spaces and places, sites and networks. These shifting alliances – which may sometimes be temporary or strategic – may be recognised as a means by which particular political aims can be achieved. As Holloway (1998a: 191) argues: 'The recognition of differences has not led to the loss of a golden age of collective action, the shifting alliances on which action is based are now simply more clear'. Thus feminist politics, revealed in practical ways in our case studies, must engage with the making of shifting alliances and recognise differences among women.

Conclusion

Our analysis in this chapter first highlighted the importance of spatiality in the construction of new femininities, and then drew out the implications of these arguments for the ways in which we think about geographies of global change and processes of identity formation. We began our analysis of the spatiality of identity formation by working through the ways in which place, at a range of different scales, matters in the constitution of new femininities. Our understanding of place is not one of bounded, distinct spaces but of places linked through a multiplicity of connections that simultaneously shape and are shaped by these wider relations, and by changing and contested relations of power and influence within these places. The femininities constituted in these places are constituted in and in turn remake these places and their connections to other places. In doing so they shape the large-scale processes of globalisation, neo-liberalism, development or transition which have been discussed throughout the book. We then argued that femininities are also formed in and through everyday spaces, understandings of which are themselves highly gendered. The dialectical and shifting nature of the relations between such sites and gender identities were explored through a consideration of the sites of paid work, home and neighbourhood.

Our second move in this chapter was to explore the implications of these arguments for current understandings of global change and changing identities.

On the one hand, our discussion highlighted the limitations of universalist discourses, the need to explore how identities work both in and across large-scale divisions of the world and the necessity of using more fluid understandings of geographical processes to examine the formation of new femininities. On the other hand, we suggested that many conceptualisations of the ways in which gender identities change also work in and through dichotomous notions of gendered space. We argued that the case studies present examples where women both work with and/or challenge these notions of gendered space, and suggested that there is a need to incorporate in our understandings of how gender identities change a more fluid conceptualisation of identity. This would take seriously the multiple intersections of gender and other identities and the shifting interconnections between gender identities and understandings of spatial categorisations and processes. In the final chapter we return to the questions with which we began the book and discuss how the femininities discussed in this book may be seen as 'new'.

CHAPTER NINE

New femininities

Introduction

In Chapter 1 we suggested that the celebration of new femininities in the British media is highly selective in its outlook. Those femininities that are celebrated are often hyped and/or transgressive. The performance of these femininities (and the emphasis on play and performance is important here) allows girls and women not only to compete with boys and men, but also to become both more and less like them. The idea of 'girl power', for example, as promoted by the British pop group the 'Spice Girls', suggests both that girls should claim power in their own right, and that this can be achieved either through hyper-feminine appearance and behaviour, or by rejecting 'traditional' versions of femininity. In straying both sides of the 'traditional' boundaries of femininity, 'girl power' is 'transgressive' and suggests that girls can be powerful without necessarily conforming to a 'traditionally' male model of behaviour. Other forms of new femininities celebrated in the media are less self-consciously transgressive, but are important as they focus on girls' and women's successes. These popular commentaries about new femininities are typically linked to a 'post-feminist' agenda which suggests that processes of diversification and fragmentation mean gender inequality is no longer an important structuring force. The implication is that girls and women can choose how to perform their femininity, asserting their difference (to men and to each other) in a world where they have equal access to resources and opportunities. Much of the media debate over such femininities has, however, centred around the 'problem' that such changes create for men's identities and life chances. Indeed, the conclusion often drawn is that new femininities reflect a reversal in power relations with women's successes being bought at the expense of men's, prompting a crisis in masculinity.

This approach to the emergence of new femininities, which focuses primarily on media-friendly transgression and on girls' and women's successes, misses much of importance. Firstly, it tends to overemphasise the 'newness' of the emergence of such changing femininities, suggesting their development is something recent and peculiar to societies experiencing late capitalism at the end of the twentieth century. Secondly, it fails to examine the importance of apparently mundane events, everyday practices, in shaping the emergence of new femininities. More importantly, it fails to consider those new femininities that are shaped by a lack of social, economic or political power, and thus glosses over the difficulties and disadvantages many women face, ignoring, for example, the widespread feminisation of poverty across different societies. Equally, the debate is centred on particular western societies and fails to take into account changing femininities in other parts of the world. In contrast, the case studies we have focused on in this book illustrate a range of ways in which women negotiate new femininities in the context of their everyday lives; how these new femininities are forged in both favourable and unfavourable circumstances, sometimes within and at other times outside the control of individual women; and how different women, living in different parts of the world, are faced with different choices in the negotiation of their gender identities. It is on these more broadly conceived new femininities that we want to focus in drawing the book to its conclusion. In the next section we examine the importance of history in the construction of new femininities, and move on in the subsequent section to look at questions of power in the reworking of femininities. We end the book with a summary of our position and the suggestion that the emergence of new femininities will frame important questions in future geographical research.

New femininities?

We argued in the introduction to the book that femininities are multiple and fractured (there is not one femininity but many and these femininities are fractured by class, ethnicity, sexuality and so on) and that these femininities are socially, historically and geographically constituted. Three examples illustrated the importance of history (as well as society and space) by showing that hegemonic femininities are not stable but change over time. We want to return to this question of history here, in order to elucidate what exactly we mean when we say *new* femininities. The new femininities described in this book are new in that these specific expressions of femininity are only just emerging, for those individuals and groups involved and/or in these particular places. However, we are not arguing that the emergence of new femininities is in itself a new phenomenon. As the historical examples in Chapter 1 demonstrate, femininities are never stabilised or fixed, but rather are subject to change within different contexts and in relation to changing historical circumstances. Thus, the production of new femininities is an ongoing process and new femininities emerge in every historical era, whether they be more or less progressive than those that went before.

This point is illustrated by the picture we chose for the cover of the book. The photograph depicts a new form of femininity in 1920s Germany, when economic, political, social and cultural changes led to a recognition of the role women could play and a celebration of their strengths and capabilities. The image of strong, fit women depicted can be seen as a forerunner to some aspects of the officially promoted femininity of communist East Germany. As Chapter 6 explained, many women in eastern Germany are now ambivalent towards this ideal. For some, the once-new femininity represented by this photograph had become reworked as a 'traditional' femininity in the GDR, one which had its problems but also its benefits. Some women now aspire to the alternative 'traditional' femininity of the 'housewife' model, the more 'conservative' form that dominated in western Germany in the post-war period. These differing 'traditions' produce differences in the feminisms associated with them and there has been much debate between eastern German and western German feminists on the desirability of particular strategies and goals in feminist politics.

Recognition that the emergence of new femininities is not a new process places 'traditional' femininities in their context. The 'traditional' femininities discussed in this book are traditional only in that they emerged at a particular moment and then became the hegemonic form of femininity in their time and place. Though 'traditional' femininities often draw on naturalising discourses and are represented as social forms which endure through the ages, they were never monolithic or fixed in time and space. Not only are 'traditional' femininities ideological constructs that do not exist in any uniform or fixed way, but also, of course, many of the femininities that might be defined as 'traditional' were in fact once new femininities, as our cover photograph suggests. These multiple 'traditional' and 'new' femininities are reworked by active contestation and by structural processes, such as labour-market restructuring.

In essence then, we are arguing that the femininities we have examined in this book are new in their recent emergence in particular places (what is 'new' in one place may be 'old' elsewhere) and for individual women in their own lives. Such developments emerge from and in turn influence the economic, political and cultural changes that constitute the uneven and unequal working out of the processes of 'globalisation' (whether it is assumed to be a process that is 'really' happening, or one powerful interpretation of current changes). The specific forms that these changes take are mediated and encouraged (sometimes demanded) by institutions and powerful discourses and ideologies associated with particular interpretations of globalisation, such as 'transition' or 'neo-liberalism', and with differing interpretations of how particular societies should operate. These interpretations are in turn themselves highly gendered (see Chapters 2 and 8). What the case studies clearly illustrate is that these processes are not in any way 'universal'. Processes of change are at one and the same time global and local, and in this context history matters in the production of new femininities. The importance of the history through which places (nations, regions, localities, and the connections within and between them) are made – shaping both local social, economic and political conditions and most particularly the locally hegemonic femininity – has already been discussed in Chapter 8. Here, we want to focus on the ways in which history matters within places, as ideas about 'traditional' femininity are redeployed and reworked in the negotiation of new femininities.

One strategy pursued by women in our case studies was to redeploy 'traditional' ideas about femininity in order to empower themselves. In Peru, as Chapters 4 and 8 suggest, women deployed dichotomous ideas about good and bad women in order to maintain their participation in the paid labour force.

'Good women' defended themselves from accusations about women behaving in a sexually immoral way by saying that they went to work not only for their children but also with them. The presence of their children in the work-place legitimated their location there and indicated that they were not 'misbehaving': they were neither abandoning their children for paid work nor likely to have sexual encounters at work because their children were accompanying them. This use of children to symbolise safety and policing in order to negotiate women's continued participation in the workfare programme has parallels with the ways in which some young Muslim women in Britain used their veils to negotiate their identities and gain access to new spaces (see Chapter 7). Here young women veiled themselves to show that they were 'good daughters' when they wanted to gain access to new spaces such as walking to school unaccompanied or planning a future away at university. Veils could also be discarded once the girls returned to 'safe' space. In both cases symbols associated with 'traditional' femininities – children and the veil – were strategically deployed and became passports to new spaces and identities. The use of particular symbols and identities enabled women and their actions to be positioned legitimately in relation to dominant discourses of gender. In so doing, the power invested in 'traditional' understandings of femininity was partially, though not wholly, subverted and new femininities began to emerge.

In saying that the power of 'traditional' femininities was partially subverted we are clearly not trying to suggest that these versions of femininity are no longer relevant. Their power is evident in that those women who do not want to live their lives by these 'traditions' need to develop a range of strategies to work around them. In this they may enjoy alliances that work across generations, in other words across what is often regarded as a key division between 'tradition' and 'progress'. In Dwyer's study some of the young British Muslim women saw their late teenage years in particular as a bounded period in their individual lives in which they could explore different femininities before more 'traditional' femities would reassert themselves as they grew older. However, other young women's mothers explicitly wanted their daughters to develop choices and experiences different from their own in life. This was by no means a clear 'rejection' of existing practices since it often occurred within a context of continued emphasis on their cultural identities as women from a British Asian background, for example. At the same time some young women reasserted a differing version of what it meant to be a young British Muslim woman by rejecting existing forms of Muslim identity as inadequate or incorrect. These represent complex, often hybrid processes by which 'new' femininities are developed.

Indeed, the other case studies also illustrate the importance of history as new femininities are constructed in relation to dominant discourses and ideologies of gender. In the reworkings of ideas about femininity there are rarely complete breaks with the past; new femininities are negotiated in relation to 'traditional' discourses, and in the process new meanings are associated with existing practices. These practices, which have been important in the definition of 'traditional' femininity, often continue to be an important and valued part of women's lives. However, they may, through collective action and female solidarity for example, be reshaped to change women's lives. In the Peruvian case study, shopping (a 'normal' part of women's reproductive responsibilities) took on a new significance for illiterate and innumerate Quechua-speaking, rural women who used their wages to shop in groups and buy in bulk. By shopping in this way, these women not only shared budgetary skills and practised entrepreneurial activities but they also learned to do business in Spanish and to recognise written numbers. This 'traditional' reproductive role therefore took on, when carried out collectively, new meanings associated with employment outside the home (the source of waged income and collective support beyond the family). Similarly, some women in Sheffield reworked traditional ideas about mothering which focused on home care to include an emphasis on paid work. This reworking of the concept of maternal responsibility in Hallam and Southey Green does not mean that the 'traditional' vision of 'the good mother' is no longer part of local culture; new understandings of maternal responsibility are defined in relation to this. The dominant ideology is not necessarily transformed – mothering or being responsible for the household remains important in the definition of femininity – but its reproduction is blurred and new, perhaps more empowering, meanings emerge.

Smith's case study of women's activism in eastern Germany suggests there are times and places where the relation to past gendered identities becomes highly problematic. For example, many women enthusiastically explore the possibilities opened up for individual and collective action as political citizens presented by the post-communist period. However, the western model of gender relations, especially in the particular context of massive economic restructuring, is reflected in the altered patterns of female labour-force participation. In addition, hegemonic western discourses typically deny the possibility of there having been positive elements to women's experiences in the GDR (for example the ability to combine family and work). Together these make it difficult to position women in Leipzig as simply drawing on or reacting against one particular tradition. Even those women who wish to adopt a 'housewife' model tend to find it an economic impossibility. In this context we find some women subordinate gendered identity to an eastern German one, emphasising solidarity across difference with eastern German men over solidarity along gender lines with women in western Germany. Other women seek to elaborate a 'different', distinctive eastern German, post-communist femininity which is not a clear rejection or adoption of any 'traditional' femininity, or even of the 'new' femininities advocated by western German feminists. In the practice of local activism, both of these strategies operate to develop distinctive, sometimes hybrid, gendered identities through what appear 'traditionally' female spaces of the home or the neighbourhood (as discussed in Chapter 8). However, appeals to western discourses of 'universal' citizenship and political rights in turn make these interpretations problematic. The nature of the negotiation of 'new' femininities is again complex and contested.

Reworking femininities

In this section we turn to an analysis of the issues of power implicated in reworking femininities. We begin by noting the range of strategies and tactics that women have used in the case studies, often with considerable success. We then stress that the extent to which women are able to access the power to implement and control such processes varies widely. Clearly, considerable divisions exist in the degree to which women have the power freely to negotiate changing gender identities or are forced into this by processes outside their individual or collective control. Finally, we return to questions of how the analysis of new femininities must deal with the fractures of gender identities and their mutual constitution with other dimensions of identity.

In Chapter 8 we have already considered some of the spatialities of the actions that women have taken in reworking their femininities. We highlighted the ways in which women used strategies of collective and collaborative support, often using the necessities or possibilities of difficult situations to develop change for themselves. We illustrated the ways in which women worked to contest the contours and assumptions of economic restructuring and challenged dominant discourses and practices such as those related to political citizenship on their own terms. Furthermore, we argued that women often work out individual solutions and accommodations within difficult situations, sometimes developing what Dwyer called 'spaces to dream', finding safe spaces in which to try out or experiment with different ways of being women. These strategies involved a range of 'informal' organisational strategies as well as developing social and political organisations and social movements, often around gender issues but also around other issues and identities, and even the adoption of 'formal' political strategies in relation to electoral power. They reshaped the gendered 'moral geographies' of, for example, mothering, work, being a teenager or being a citizen. Identifying the ways in which women have actively negotiated and shaped new femininities helps focus analysis on the everyday ways in which women mobilise multiple representations of themselves, often juggling a series of difficult and demanding roles. This they do while planning a better life for themselves, and often for their families and neighbourhoods. This not only highlights the mechanisms through which existing femininities are reworked but also makes these often overlooked forms of change a focus for analysis.

Women's actions involve both positions of strength and of weakness, often combined in ambiguous ways. The question of power is therefore an important one. Anything other than crass generalisation of the case studies shows clearly that different women have different access to the resources, images and symbols necessary to contest and rework hegemonic femininities. Some girls and women lack the power to redeploy 'traditional' understandings of femininity in new ways or to attach new meanings to 'traditional' practices, let alone to reject outright the hegemonic femininity of their time and place. For example, for young women in the two schools in 'Hertfield' their teenage years and the spaces created with friends and at school provided opportunities to experiment with what it meant for them to be young British Muslim women, sometimes literally dressing-up and performing different femininities. However, while the middle-class girls could draw on networks of support and encouragement from teachers and parents to imagine a range of possible future paths, the expectations and encouragement for working-class girls were much less in evidence. For these young women there are then considerable divergences in access to repertoires of resources and subjectivities on which they can draw to develop a diversity of new femininities. The 'performance' of new femininities is not totally voluntaristic: some women simply cannot make these choices and there are different costs to those who are able, in some way or another, to do so. For low-income women in Peru the space to explore new femininities is bound up with finding a space of economic security through paid work in a time of severe economic crisis, and for many eastern German women the space to explore new political gendered identities comes with uncertainty in employment and housing. In these spaces identities are not just imagined but are also performed in actions which can sometimes seem like play-acting and experimentation and at other times reflect very real, everyday experiences which may be tied to issues as fundamental as material survival.

It is important to be clear, then, that some women negotiate new femininities not through choice but through force of circumstance. New femininities are not only shaped by women who have been able to mobilise enough power to transgress the boundaries of 'traditional' femininity, they are often shaped by changing social and economic structures outside the control of individual women. The Sheffield case study illustrates both sides of this issue well. Some of the mothers interviewed by Holloway actively wanted to renegotiate 'traditional' understandings of mothering which were centred around home care to include an emphasis on mothers' participation in paid work, and some of these (though importantly not all) were able to mobilise sufficient cultural and material resources to be able to do so. However, other women, who might have preferred to stay at home for some time with their children, were obliged by labour market structures to go out to work – for example, women who held relatively skilled jobs risked marginalisation in the labour market if they opted to stay at home while their children were young. Social and economic structures thus shaped their decisions. Changing identities are therefore in many cases bound up with real, everyday experiences of material production and reproduction. However, it is also evident that even where the need for 'survival strategies' is less obvious, identity formation is embedded in and structured by the social and economic differences discussed in Chapters 2 and 8. Nevertheless, while not wishing to overemphasise this point, the case studies also point to evidence that changes in these structures may (often inadvertently) produce new opportunities for women to renegotiate their individual and collective femininities. Furthermore, the ways in which women work to create such change in turn serve to shape the very process of these changes, albeit in very diverse ways.

This focus on power is essential, since in discussing the reworking of femininities it is important not to represent women and girls as endlessly inventive and flexible because the essentialising power of such images can represent female resilience as something almost 'natural' rather than as socially constructed and in many cases as something that is bought at a great cost. Moser (1993) makes a similar point in her discussion of the over-celebration of women's deployment of survival strategies in times of economic restructuring. She cautions that, while celebratory approaches avoid conceptualising women merely as victims of change, they also ignore the fact that many women experience great stress while trying to perform multiple roles, as some of the comments on the limitations of local activism among women in Leipzig showed. This point is crucial here because while some women are able to rework femininities with relative ease, for others this reworking is bought at emotional and social cost to themselves and their loved ones, and in extreme cases through the risk or reality of physical violence. While these difficulties often operate through immediate personal circumstances and relationships, they are also shaped by the differential effects of processes of globalisation, cultural difference and political restructuring and by international, national, regional and local geographies of class, generation, racism, heterosexism and patriarchy.

Clearly, the strategies and opportunities to rework female identities are not without their contradictions and are, moreover, not automatically 'progressive' for all women. For example, women themselves may redeploy 'traditional' ideas about 'good' and 'bad' women in the negotiation of new femininities discussed above. These understandings of 'good' and 'bad' women, mobilised for example in the PAIT programme, are not only constructed in relation to hegemonic gender identities but they are also dichotomous, that is they are defined in relation to one another. Thus women in the PAIT programme did not simply define themselves as good. Instead they showed how different they were from 'bad' women, often using physical space to distance themselves from the 'bad' women. However the parameters of what is 'good' and 'bad' and the relationship between them were fluid and open to negotiation. Particularly important was the audience for whom the representation of 'good' and 'bad' was being made. 'Good women' in PAIT might emphasise their difference from the 'bad women' to their husbands/partners, but in other circumstances, for example with their friends, they expressed more solidarity with these women, emphasising that single mothers were not 'all bad' as they worked very hard to care for their children. What was 'good' shifted according to what was 'bad', and this was strongly influenced by the audience or context for that representation.

The importance of audience or social context for the negotiation of feminine identities can also be seen in Dwyer's work. The young Muslim women she interviewed attempted to undermine their non-Asian friends' preconceived ideas about arranged marriages and oppressive family relations at school. At home they tried to resist social control while showing their families that they were still trustworthy, good daughters. These young women confront ideas about femininity at home and school that, at different times, they both contest and reinforce. The categories of 'good' and 'bad' women, 'good' and 'bad' behaviour are not, then, simply given absolutes. They are mobilised in different and sometimes contradictory ways at different times and in different places. Consequently, the subjectivities that women occupy in the construction of new femininities are not necessarily internally consistent. To some extent at least, new femininities are forged in the spaces created by the variety of sometimes contradictory representations made in the course of everyday resistance to different sets of ideas about appropriate gender behaviour. Some of these ideas are more normalising than others because they are embodied in people and institutions with varying degrees and types of power.

Differences among women and the contradictory and complex ways women represent themselves and others serve to emphasise that the transgression of the boundaries of 'traditional' femininities and the emergence of new ones is not necessarily progressive. Indeed, the interpretation of transgression depends very much on the audience, and this includes different groups of women. For example, across the case studies while some women may feel that paid work outside the home is a benefit, others may see it as a retrograde step for children, women and society. To state that the interpretation of transgression depends on the context is not to slip into relativism. Rather, in exploring the emergence of new femininities, we would argue that individuals and groups of women need to be empowered to make their own choices about femininity. The purpose of feminism as a political movement is not to reject all 'traditional' understandings of femininity and suggest one single 'more enlightened' version; rather it is to empower a diversity of women to make their own choices about their lives, recognising that the choices that women make will differ.

The geographies of new femininities

Our aim in this volume has been to highlight and explore, through an explicitly geographical approach, the emergence of new femininities. In pursuing this agenda we have usefully linked together debates about globalisation and work on identity, literatures that are all too often considered in isolation. The economic, political and cultural processes associated with broad processes of 'globalisation', 'neo-liberalism', 'transition' or 'development' are without doubt key influences in shaping new femininities. These processes not only reshape the relative position of different countries, regions and localities around the world, but the very processes themselves are gendered, affecting men and women differently. However, we have been careful to try to avoid a determinist argument which suggests that these global processes are all-powerful and directly shape new femininities. It is equally clear that place matters in the sense that existing national and sub-national histories and responses are important. More particularly, we have also been careful to elucidate the ways in which differences among women shape the emergence of new femininities. The different subject positions women occupy, through different class, ethnic, age and religious backgrounds, and so on, are crucial in the construction of their gender identities, and in the ways they experience and respond to these processes of social change. Moreover, this focus on the specificity of identity formation illustrates the importance of social practices in, and ideas about, everyday spaces as well as wider regional, national or global links in the emergence of new femininities. These are mediated in and through institutions such as international financial and political organisations, through the gendered geographies of the national, regional and local state and through moral geographies of a range of spaces such as the household, the neighbourhood, the school or the work-place.

Thus our analysis shows that processes involved in globalisation and in identity formation are closely bound together, and are, moreover, spatially as well as socially constituted. In particular, our focus on the social and spatial constitution of processes of globalisation and identity formation has been explored through our assertion that space matters in three interlinked ways. Place matters as the context for the negotiation of new femininities; the social relations of everyday spaces matter in the construction and contestation of new femininities; and gendered discourses matter as they shape the meanings of everyday spaces. These understandings of gender and spatiality informed our research methodologies, were developed empirically through our analysis of the four individual empirical case studies which we have sought to link together and analyse collectively in this book, and in turn were reflected in our re-engagement with macro-theories of social change and understandings of changing gender identities.

We considered how the methodological issues involved in working with 'gender' and with 'geography' raised a variety of questions about the ways in which research on changing femininities and their geographies might deal explicitly with the questions of feminist epistemology around issues of power and positionality. We also gave a range of examples of how theoretical questions around the links between gender and geography and about the fracturing of identity might inform the research practice and in turn influence the development of our theories. Thus through the case studies and in the comparative analysis of Chapter 8, we highlighted the ways in which an engagement with new femininities enriched and disrupted several key discourses of macro-level change. Furthermore, the book as a whole demonstrates the varied and often complex intersections, in different spaces, sites and places, of gender with other identities and illustrates the fractured, fracturing and changing nature of femininities. Overall, the book demonstrates the continued importance of studies that work with fractured and fracturing understandings of gender.

Our focus on the geographies of new femininities in this book opens up important questions for future geographical research. Firstly, we have reiterated the extent to which geography and gender cannot be thought about separately. Instead, as we have demonstrated, social and spatial processes are inextricably linked together through the gendering of spaces and the making of identities through places, sites and networks. We have also emphasised that complex economic, political and cultural processes such as globalisation can be more fully understood by adopting an approach such as ours which considers the gendered consequences and contexts of such processes beyond the macro level. As we have argued, the approach that we have adopted in this book has allowed us to critique existing theoretical understandings of global processes by our close attention to the ways in which such processes are effective and have meaning at the scale of the everyday lives of ordinary women in different parts of the world. Secondly, our approach in this book raises important questions about the theorisation of gender and feminist agendas. A key focus of our work has been to highlight differences among women, both within and between places. We have sought to demonstrate the ways in which gender is a fractured social identity which is mutually constituted with other identities and which, fundamentally, shifts and changes over time and space. However, at the same time, one of the interesting findings of our shared analysis has been the resonances and similarities between womens' lives in very different contexts. This tension highlights current dilemmas for a feminist politics which must engage with this issue of difference (Scott *et al.*, 1997). For feminist geographers our work both raises questions about the theorisation of gender as a founding social division and suggests some of the complexities of working with gender as a category. Our conclusion emphasises the extent to which gender remains an important social category but must often be understood as one identity within a context of shifting alliances rather than as one shared essential identity.

This book has been innovative in its use of four different case studies through which a shared analysis has been developed. Our argument has been that it is through this kind of cross-cultural research that important insights about the construction and contestation of 'new' femininities are made possible. Through our analysis of four very different case studies we have been able to draw out both generalisations and specificities about the changing lives of women and the meanings of 'new' femininities. Our comparative methodology also compelled us constantly to question our own individual standpoints in relation to our empirical work as we worked towards writing this book. This cross-cultural approach is, we believe, one of the book's strengths. Our comparative approach has made it clear that analysis of femininities must always engage with questions of place, context and history. If we return to a final reconsideration of the cover image of this book, we are reminded that representations of femininities are read in different ways at different times and so can have a series of new meanings attached to them. Now, seven decades later, for example, the black and white photography of this image has an almost sensual appeal for viewers more familiar with full colour photography. This sensuality may or may not have been part of the original idea behind the photograph when it was first produced and what is sensual in one context does not necessarily translate into another time and place. The fact that we cannot be sure of the photographer's and the subjects' intentions exemplifies an important point, which is that not only do femininities change but also it is possible to attribute different and sometimes new meanings to them across time and space.

Further reading

The readings listed below indicate some key readings for those interested in following up on some of the areas of concern to the four case studies in this book.

Emergency work and 'workfare'

BILLIONE, J. (1986) *El PAIT Funcionalidad y Metologías*, COOPOP, Lima.

EMPLOYMENT COMMITTEE, HOUSE OF COMMONS (1996) *The Right to Work/ Workfare* (Second Report), HMSO, London.

GAUDE, J. and WATZLAWICK, H. (1992) Employment creation and poverty alleviation through labour intensive public works in least developed countries, *International Labour Review*, **131**, 3–18.

GRAHAM, C. (1991) Urban poor: the PAIT programme in Lima's Pueblos Jovenes, *Journal of Latin American Studies*, **23**, 91–130.

GRAHAM, C. (1992) The politics of protecting the poor during adjustment: Bolivia's Emergency Social Fund, *World Development*, **20**, 1233–1251.

GUERON, J. and PAULY, E. (1991) *From Welfare to Work*, Russell Sage Foundation, New York.

PAREDES, P. (1988) Nadie sabe para quien trabaja: la mujer y el PAIT, in Paredes, P. and Tello, G. (eds) *Pobreza Urbana y Trabajo Feminino*, ADEC and ATC, Lima, pp. 71–242.

PECK, J. (1998) Workfare in the sun: politics, representation, and method in US welfare-to-work strategies, *Political Geography*, **17**, 535–566.

PECK, J. (1998) Workfare: a geopolitical etymology, *Environment and Planning D: Society and Space*, **16**, 133–161.

PREALC (1988) *Empleos de Emergencia*, OIT, Santiago.

WURGAFT, J. (1992) Social investment funds and economic restructuring in Latin America, *International Labour Review*, **131**, 35–44.

Peru: political economy

CAMERON, M. and MAUCERI, P. (eds) (1997) *The Peruvian Labyrinth: Polity, Society, Economy*, The Pennsylvania State University Press, University Park, PA.

CRABTREE, J. (1992) *Peru Under García: An Opportunity Lost*, Macmillan, Basingstoke.

GLEWWE, P. and DETRAY, P. (1991) The poor in Latin America during adjustment: a case study of Peru, *Economic Development and Cultural Change*, **4**, 27–54.
GRAHAM, C. (1992) *Peru's APRA, Parties, Politics and the Elusive Quest for Democracy*, Lynne Rienner, London.
POOLE, D. and RENIQUE, G. (1992) *Peru: Time of Fear*, Latin American Bureau, London.
RUDOLPH, J. (1991) *Peru: the Evolution of a Crisis*, Stanford University Press, Stanford, CA.
STARN, O., DEGREGORI, C. and KIRK, R. (eds) (1995) *The Peru Reader*, Duke University Press, Durham, NC.
TANSKI, J. (1994) The impact of crisis, stablization and structural adjustment on women in Lima, Peru, *World Development*, **22**, 1627–1642.

Gender identities in Latin America

ACOSTA-BELÉN, E. and BOSSE, C. (eds) (1993) *Researching Women in Latin America and the Caribbean*, Westview Press, Boulder, CO.
BABB, F. (1990) Women and work in Latin America, *Latin American Research Review*, **25**, 236–247.
FISHER, J. (1993) *Out of the Shadows. Women, Resistance and Politics in South America*, Latin American Bureau, London.
JELIN, E. (ed.) (1990) *Women and Social Change in Latin America*, Zed Books, London.
KUPPERS, G. (1994) *Compañeras. Voices from the Latin American Women's Movement*, Latin American Bureau, London.
LIND, A. (1992) Power, gender, and development: popular women's organizations and the politics of needs in Ecuador, in Escobar, A. and Alvarez, S. (eds) *The Making of Social Movements in Latin America: Identity, Strategy, and Democracy*, Westview, Oxford, pp. 134–149.
MELHUUS, M. and STOLEN, K. (eds) (1996) *Machos, Mistresses, Madonnas. Contesting the Power of Latin American Gender Imagery*, Verso, London.
RADCLIFFE, S. and WESTWOOD, S. (eds) (1993) *'Viva' Women and Popular Protest in Latin America*, Routledge, London.
STEPHEN, L. (1997) *Women and Social Movements in Latin America. Power from Below*, University of Texas Press, Austin, TX.
TOWNSEND, J. (1995) *Women's Voices from the Rainforest*, Routledge, London.
WAYLAND, G. (1993) Women's movements and democratisation in Latin America, *Third World Quarterly*, **14**, 573–587.

Motherhood and mothering

BARTLETT, J. (1994) *Will You Be Mother?: Women Who Choose To Say No*, Virago, London.
BELL, L. (1995) Just a token commitment?: women's involvement in a local baby-sitting circle, *Women's Studies International Forum*, **18**, 325–336.
BELL, L. and RIBBENS, J. (1994) Isolated housewives and complex maternal worlds: the significance of social contacts between women with young children in industrial societies, *Sociological Review*, **42**, 227–262.
DUNNE, G.A. (1998) 'Pioneers behind our own front doors': towards greater balance in the organisation of work in partnerships, *Work, Employment and Society*, **12**, 273–295.

DYCK, I. (1990) Space, time and renegotiating motherhood: an exploration of the domestic workplace, *Environment and Planning D: Society and Space*, **8**, 459–483.

ENGLAND, K. (1995) Girls in the office – recruiting and job search in a local clerical labor-market, *Environment and Planning A*, **27**, 1995–2018.

HANSON, S. and PRATT, G. (1995) *Gender, Work and Space*, Routledge, London.

KITZINGER, S. (1993) *Ourselves as Mothers: The Universal Experience of Motherhood*, Bantam Books, London.

PRATT, G. and HANSON, S. (1991) On the links between home and work: family-household strategies in a buoyant labor-market, *International Journal of Urban and Regional Research*, **15**, 55–74.

WARD, C., DALE, A. and JOSHI, H. (1996) Combining employment and childcare: an escape from dependence?, *Journal of Social Policy*, **25**, 223–247.

Childcare

BROPHY, J. (1994) Parent management committees and pre-school playgroups: the partnership model and future management policy, *Journal of Social Policy*, **23**, 161–194.

BROPHY, J., STATHAM, J. and MOSS, P. (1992) *Playgroups in Practice: Self-help and Public Policy*, HMSO, London.

COTTERILL, P. (1992) 'But for freedom, you see, not to be a babyminder': women's attitudes towards grandmother care, *Sociology*, **26**, 603–618.

ENGLAND, K. (ed.) (1996) *Who Will Mind The Baby? Geographies of Childcare and Working Mothers*, Routledge, London.

FINCHER, R. (1991) Caring for workers' dependants: gender, class and local state practice in Melbourne, *Political Geography Quarterly*, **10**, 356–381.

MACKENZIE, S. and TRUELOVE, M. (1993) Changing access to public and private services: non-family childcare, in Bourne, L.S. and Ley, D.F. (eds) *The Changing Social Geography of Canadian Cities*, McGill-Queen's University Press, Montreal, pp. 326–342.

PRATT, G. (1997) Stereotypes and ambivalence: the construction of domestic workers in Vancouver, British Columbia, *Gender, Place and Culture*, **4**, 159–177.

ROSE, D. and CHICOINE, N. (1991) Access to school daycare services: class, family, ethnicity and space in Montréal's old and new inner city, *Geoforum*, **22**, 185–201.

ROSE, D. (1993) Local childcare strategies in Montréal, Quebec: the mediations of state policies, class and ethnicity in the life courses of families with young children, in Katz, C. and Monk, J. (eds) *Full Circles: Geographies of Women over the Life Course*, Routledge, London, pp. 188–207.

SMITH, F.M.B. and BARKER, J. (forthcoming) From Ninja Turtles to the Spice Girls: childrens' participation in the development of out-of-school environments, *Built Environment*.

TRUELOVE, M. (1993) Measurement of spatial equality, *Environment and Planning C: Government and Policy*, **11**, 19–34.

Women in eastern Europe and the post-Soviet states

BRIDGER, S., KAY, R. and PINNICK, K. (1996) *No More Heroines? Russia, Women and the Market*, Routledge, London.

BUCKLEY, M. (ed.) (1997) *Post-Soviet Women: From the Baltic to Central Asia*, Cambridge University Press, Cambridge.
CORRIN, C. (ed.) (1992) *Superwomen and the Double Burden*, Scarlet Press, London.
EINHORN, B. (1993) *Cinderella Goes to Market: Citizenship, Gender and Women's Movements in East Central Europe*, Verso, New York.
FUNK, N. and MUELLER, M. (eds) (1993) *Gender Politics and Post-communism*, Routledge, London.

Women in the GDR and German reunification

BÜTOW, B. and STECKER, H. (eds) (1994) *EigenArtige Ostfrauen*, Kleine Verlag, Bielefeld.
DÖLLING, I. (1989) Continuity and change in the media image of women in the GDR, in Gerber, M. and Woods, R. (eds) *Studies in GDR Culture and Society 8*, University Press of America, Lanham, MD, pp. 131–144.
FERREE, M.M. (1997) German unification and feminist identity, in Scott, J.W., Kaplan, C. and Keates, D. (eds) *Transitions, Environments, Translations: Feminisms in International Politics*, Routledge, New York, pp. 46–55.
HAMPELE, A. (1993) The organised women's movement in the collapse of the GDR: the Independent Women's Association (UFV), in Funk, N. and Mueller, M. (eds) *Gender Politics and Post-communism*, Routledge, London, pp. 194–200.
KOLINSKY, E. (1993) *Women in Contemporary Germany: Life, Work and Politics*, Providence, Oxford.
ROHNSTOCK, K. (ed.) (1994) *Stiefschwestern: Was Ost-Frauen und West-Frauen voneinander denken*, Fischer, Frankfurt.
SCHMUDE, J. (1996) Contrasting developments in female labour force participation in East and West Germany since 1945, in García-Ramon, M.D. and Monk, J. (eds) *Women of the European Union*, Routledge, London, pp. 156–185.
WINKLER, G. (ed.) (1990) *Frauenreport '90*, Verlag Die Wirtschaft, Berlin.

German reunification

BLACKSELL, M. and SMITH, F.M. (eds) (1997) Special Issue: Aspects of German Unification, *Applied Geography*, **17**, 255–396.
BLACKSELL, M., BORN, K.M. and BOHLANDER, M. (1996) Settlement of property claims in former East Germany, *Geographical Review*, **86**, 199–215.
JONES, A. (1994) *The New Germany: A Human Geography*, Wiley, Chichester.
OSMOND, J. (ed.) (1992) *German Reunification: A Reference Guide and Commentary*, Longman, Harlow.
OSTERLAND, M. (1994) Coping with democracy: the re-institution of self-government in Eastern Germany, *European Urban and Regional Research*, **1**, 5–18.

Femininity and adolescent girls

LEES, S. (1993) *Sugar and Spice: Sexuality and Adolescent Girls*, Penguin, London.
McROBBIE, A. (1991) *Feminism and Youth Culture*, Macmillan, London.
SKEGGS, B. (1997) *Formations of Class and Gender*, Sage, London.

The theorisation of 'new ethnicities'

BACK, L. (1996) *New Ethnicities and Urban Culture: Racisms and Multiculture in Young Lives*, UCL Press, London.
HALL, S. (1992) New ethnicities, in Donald, J. and Rattansi, A. (eds) *'Race', Culture and Difference*, pp. 252–259. Originally published in Mercer, K. (ed.) (1988) *Black Film, British Cinema*, Institute of Contemporary Arts, London, pp. 27–31.

Islam in Britain

LEWIS, P. (1994) *Islamic Britain*, I. B. Taurus, London.
VERTOVEC, S. and PEACH, C. (1997) *Islam in Europe: The Politics of Religion and Community*, Macmillan, London.

Identities of young British South Asian women

BHACHU, P. (1993) Identities constructed and reconstructed: representations of Asian women in Britain, in Buijs, G. (ed.) *Migrant Women: Crossing Boundaries and Changing Identities*, Berg, Oxford, pp. 99–117.
BRAH, A. (1993) 'Race' and 'culture' in the gendering of labour markets: South Asian young Muslim women and the labour market, *New Community*, **19**, 441–458.
KASSAM, N. (1997) *Telling It Like It Is: Young Asian Women Talk*, The Women's Press, London.
KNOTT, K. and KHOKHER, S. (1993) Religion and ethnic identity among young Muslim women in Bradford, *New Community*, **19**, 593–610.

Debates about the veil

ABU ODEH, L. (1993) Post-colonial feminism and the veil: thinking the difference, *Feminist Review*, **43**, 26–37.
ALI, Y. (1992) Muslim women and the politics of ethnicity and culture in Northern England, in Sahgal, G. and Yuval-Davis, N. (eds) *Refusing Holy Orders*, Virago, London, pp. 101–123.
ALIBHAI-BROWN, Y. (1994) Sex, veils and stereotypes, *The Independent*, 22 December, p. 17.

Bibliography

ABU ODEH, L. (1993) Post-colonial feminism and the veil: thinking the difference, *Feminist Review*, **43**, 26–37.

AFSHAR, H. (1989a) Gender roles and the 'moral economy of kin' among Pakistani women in West Yorkshire, *New Community*, **15**, 211–225.

AFSHAR, H. (1989b) Education: hopes, expectations and achievements of Muslim women in West Yorkshire, *Gender and Education*, **1**, 261–272.

AFSHAR, H. (1994) Muslim women in West Yorkshire: growing up with real and imaginary values amidst conflicting views of self and society, in Afshar, H. and Maynard, M. (eds) *The Dynamics of 'Race' and Gender: Some Feminist Interventions*, Taylor and Francis, London, pp. 127–150.

AFSHAR, H. (ed.) (1996) *Women and Politics in the Third World*, Routledge, London.

AGGER, I. (1994) *The Blue Room: Trauma and Testimony Among Refugee Women. A Psycho-social Exploration*, Zed Press, London.

AHMED, L. (1992) *Women and Gender in Islam*, Yale University Press, London.

ALI, Y. (1992) Muslim women and the politics of ethnicity and culture in Northern England, in Sahgal, G. and Yuval-Davis, N. (eds) *Refusing Holy Orders*, Virago, London, pp. 101–123.

ALIBHAI-BROWN, Y. (1994) Sex, veils and stereotypes, *The Independent*, 22 December, p. 17.

ALLEN, T. and HAMNETT, C. (eds) (1995) *A Shrinking World? Global Unevenness and Inequality*, Oxford University Press, Oxford.

ALTVATER, E. (1993) *The Future of the Market*, Verso, London.

ALVAREZ, S., DAGNINO, E. and ESCOBAR, A. (eds) (1998) *Cultures of Politics and Politics of Cultures*, Westview Press, Boulder, CO.

AMIN, A. (1997) Globalisation: an 'in here-out there' connectivity, paper presented at the 'After Globalisation' conference, Reading University, 7 May.

AMIN, A. and TOMMANEY, J. (eds) (1995) *Behind the Myth of European Union: Prospects for Cohesion*, Routledge, London.

ANDERSON, J. (1995) The exaggerated death of the nation-state, in Anderson, J., Brook, C. and Cochrane, A. (eds) *A Global World: Re-ordering Political Space*, Oxford University Press, Oxford, pp. 65–112.

APPIGNANESI, L. and MAITLAND, S. (1989) *The Rushdie File*, Fourth Estate, London.

ARREGUI, M. and BAEZ, C. (1991) Free trade zones and women workers, in Wallace, T. and March, C. (eds) *Changing Perceptions: Writings on Gender and Development*, Oxfam, Oxford, pp. 30–39.

BACK, L. (1996) *New Ethnicities and Urban Culture: Racisms and Multiculture in Young Lives*, UCL Press, London.

BARRIG, M. (1989) The difficult equilibrium between bread and roses: women's organisations and the transition from dictatorship to democracy in Peru, in Jaquette, J. (ed.) *The Women's Movement in Latin America and the Transition to Democracy*, Unwin Hyman, Boston, MA, pp. 114–148.

BARTLETT, J. (1994) *Will You Be Mother?: Women Who Choose to Say No*, Virago, London.

BARTON, J. (1997) *A Political Geography of Latin America*, Routledge, London.

BEAVERSTOCK, J.V. (1996a) Revisiting high-waged labour market demand in the global cities: British professional and managerial workers in New York City, *International Journal of Urban and Regional Research*, 20, 422–445.

BEAVERSTOCK, J.V. (1996b) Migration, knowledge and social interaction: expatriate labour within investment banks, *Area*, 28, 459–470.

BELL, D. and VALENTINE, G. (eds) (1995) *Mapping Desire: Geographies of Sexualities*, Routledge, London.

BELL, D., CAPLAN, P. and KARIM, W. (eds) (1993) *Gendered Fields. Women, Men and Ethnography*, Routledge, London.

BELL, D., BINNIE, J., CREAM, J. and VALENTINE, G. (1994) All hyped up and no place to go, *Gender, Place and Culture*, 1, 31–48.

BELL, L. and RIBBENS, J. (1994) Isolated housewives and complex maternal worlds: the significance of social contacts between women with young children in industrial societies, *Sociological Review*, 42, 227–262.

BENHABIB, S. (1992) Models of public space: Hannah Arendt, the Liberal tradition and Jürgen Habermas, in Calhoun, C. (ed.) *Habermas and the Public Sphere*, MIT Press, Cambridge, MA, pp. 73–98.

BENN, M. (1998) *Towards a New Politics of Motherhood*, Jonathan Cape, London.

BERNEDO, J. (1989) *PAIT: Fundamentos, Procesos y Opciones*. Fundación Friedrich Ebert, Lima.

BHABHA, H. (1990) The third space: an interview with Homi Bhabha, in Rutherford, J. (ed.) *Identity, Community, Culture, Difference*, Lawrence and Wishart, London, pp. 207–221.

BHABHA, H. (1992) Postcolonial authority and postcolonial guilt, in Grossberg, L., Neilson, C. and Treichler, P. (eds) *Cultural Studies*, Routledge, New York, pp. 69–80.

BHACHU, P. (1993) Identities constructed and reconstructed: representations of Asian women in Britain, in Buijs, G. (ed.) *Migrant Women: Crossing Boundaries and Changing Identities*, Berg, Oxford, pp. 99–117.

BHOPAL, K. (1997) *Gender, Race and Patriarchy: A Study of South Asian Women*, Ashgate, London.

BLACKSELL, M., BORN, K.M. and BOHLANDER, M. (1996) Settlement of property claims in former East Germany, *Geographical Review*, 86, 199–215.

BÖCKMANN-SCHEWE, L., KULKE, C. and RÖHRIG, A. (1995) 'Es war immer so, den goldenen Mittelweg zu finden zwischen Familie und Beruf war eigentlich das Entscheidende' Kontinuitäten und Veränderungen von Frauen in den neuen Bundesländern, *Berliner Journal für Soziologie*, 2, 207–222.

BONDI, L. (1998) Sexing the city, in Jacobs, J. and Fincher, R. (eds) *Cities of Difference*, Guilford Press, New York, pp. 177–200.

BORNEMAN, J. (1992) *Belonging in the Two Berlins: Kin, State, Nation*, Cambridge University Press, Cambridge.

BRAH, A. (1993) 'Race' and 'culture' in the gendering of labour markets: South Asian young Muslim women and the labour market, *New Community*, **19**, 441–458.

BRAH, A. and MINHAS, R. (1985) Structural racism or cultural difference: schooling for Asian girls, in Weiner, G. (ed.) *Just a Bunch of Girls*, Open University Press, Milton Keynes, pp. 14–25.

BRAND, D. (1990) Bread out of stone, in Scheier, L., Sheard, S. and Wachtd, E. (eds) *Language in Her Eye: Views on Writing and Gender by Canadian Women Writing in English*, Coach House Press, Toronto, pp. 45–53.

BRANNEN, J. and MOSS, P. (1988) *New Mothers at Work*, Unwin, London.

BRIDGER, S., KAY, R. and PINNICK, K. (1996) *No More Heroines? Russia, Women and the Market*, Routledge, London.

BROOK, C. (1995) The drive to global regions?, in Anderson, J., Brook, C. and Cochrane, A. (eds) *A Global World: Re-ordering Political Space*, Oxford University Press, Oxford, pp. 113–166.

BROPHY, J., STATHAM, J. and MOSS, P. (1992) *Playgroups in Practice: Self-help and Public Policy*, HMSO, London.

BROWNHILL, S. and HALFORD, S. (1990) Understanding women's involvement in local politics: how useful is a formal/informal dichotomy?, *Political Geography Quarterly*, **9**, 396–414.

BRUNO, M. (1995) The second love of worker bees: gender, employment and social change in contemporary Moscow, paper presented at the Annual Conference of the Institute of British Geographers, January.

BRUNO, M. (1997) Women and the culture of entrepreneurship, in Buckley, M. (ed.) *Post-Soviet Women: from the Baltic to Central Asia*, Cambridge University Press, Cambridge, pp. 56–74.

BRUNT, R. (1990) The politics of identity, in Hall, S. and Jacques, M. (eds) *New Times: the Changing Face of Politics in the 1990s*, Lawrence and Wishart (in association with *Marxism Today*), London, pp. 150–159.

BURGESS, J. (1998) 'But is it worth the risk?' How women negotiate access to urban woodland: a case study, in Ainley, R. (ed.) *New Frontiers of Space, Bodies and Gender*, Routledge, London, pp. 115–128.

BURGESS, J., LIMB, B. and HARRISON, C. (1988a) Exploring environmental values through the medium of small groups. Part One: theory and practice, *Environment and Planning A*, **20**, 309–326.

BURGESS, J., LIMB, B. and HARRISON, C. (1988b) Exploring environmental values through the medium of small groups. Part Two: illustrations of a group at work, *Environment and Planning A*, **20**, 457–476.

BURGOS DEBRAY, E. (ed.) (1984) *I, Rigoberta Menchú, An Indian Woman in Guatemala*, Verso Press, London.

BUTLER, J. (1990) *Gender Trouble*, Routledge, London.

BUTLER, J. (1993) *Bodies That Matter: On the Discursive Limits of 'Sex'*, Routledge, New York.

BÜTOW, B. (1994) Politische Nichtpartizipation von Frauen?, in Bütow, B. and Stecker, H. (eds) (1994) *EigenArtige Ostfrauen*, Kleine Verlag, Bielefeld, pp. 261–268.

BÜTOW, B. and STECKER, H. (1994) Statt eines Nachwortes, in Bütow, B. and Stecker, H. (eds) (1994) *EigenArtige Ostfrauen*, Kleine Verlag, Bielefeld, pp. 324–326.

Canadian Geographer (1993) Feminism as method, Special Issue, Moss, P. (ed.), **37**, 48–61.

CANEL, E. (1997) New social movements theory and resource mobilisation theory: the need for integration, in Kaufman, M. and Alfonso, H. (eds) *Community Power and Grassroots Democracy*, Zed Press, London, pp. 189–222.

CARAWAY, N. (1991) *Segregated Sisterhood. Racism and the Politics of American Feminism*, University of Tennessee Press, TN.

CASSELL, M.K. (1996) The Treuhandanstalt, privatization and the role of the courts, *Discussion Paper FS 1 96–316*, Wissenschaftszentrum Berlin für Sozialforschung, Berlin.

CASTELLS, M. (1997) *The Power of Identity*, Blackwell, London.

CASTLES, S. and MILLER, M. (1993) *The Age of Migration*, Macmillan, Basingstoke.

CHAMPION, A.G. and TOWNSEND, A.R. (1990) *Contemporary Britain: A Geographical Perspective*, Edward Arnold, London.

CHANEY, E. (1979) *Supermadre: Women in Politics in Latin America*, Austin University Press, Austin, TX.

CHANT, S. (1997) *Women-headed Households: Diversity and Dynamics in the Developing World*, Macmillan, Basingstoke.

CHOUINARD, V. and GRANT, A. (1995) On being not even anywhere near the project – ways of putting ourselves in the picture, *Antipode*, **27**, 137–166.

CLARK, F. and LAURIE, N. (forthcoming) Old age, gender and marginality in Peru: a state in transition, *Bulletin of Latin American Research*.

COCHRAN, M. and GUNNARSSON, L. with GRÄBE, S. and LEWIS, J. (1993) The social networks of coupled mothers in four cultures, in Cochran, M., Larner, M., Rily, D., Gunnarsson, L. and Henderson, C.R. Jr (eds) *Extending Families: The Social Networks of Parents and their Children*, Cambridge University Press, Cambridge, pp. 86–104.

COCKS, J. (1989) *The Oppositional Imagination: Feminism Critique and Political Theory*, Routledge, London.

COHEN, B. (1988) *Caring for Children: Services and Policies for Childcare and Equal Opportunities in the United Kingdom*. Report for the European Commission's Childcare Network, Family Policy Studies Centre, London.

CONNELL, R.W. (1995) *Masculinities*, Polity Press, Cambridge.

CONTRERAS IVARCENA, E. (1991) *La Violencia Política en Apurímac su Impacto Social y Económico*, Centro de Estudios Regional Andinos 'Bartolomé de las Casas', Cusco.

CORRIN, C. (ed.) (1992) *Superwomen and the Double Burden*, Scarlet Press, London.

COULSON, J. (forthcoming) Neo-liberalism and gender in Ecuador: the case of the flower industry, Unpublished PhD Thesis, Newcastle University, UK.

CPU (Central Policy Unit) (1993) *Sheffield 1991 Census Report 4: Summary Ward Profiles for Sheffield*, Sheffield City Council, Sheffield.

CRANG, P. (1992) The politics of polyphony: reconfigurations in geographical authority, *Environment and Planning D: Society and Space*, **10**, 527–549.

CRC (Community Relations Commission) (1976) *Between Two Cultures: A Study of Relationships Between Generations in the Asian Community in Britain Today*, CRC, London.

CREAM, J. (1995) Women on trial: a private pillory?, in Pile S. and Thrift, N. (eds) *Mapping the Subject: Geographies of Cultural Transformation*, Routledge, London, pp. 158–169.

CRUSH, J. (1995) *Power of Development*, Routledge, London.

DALE, J. and FOSTER, P. (1986) *Feminists and State Welfare*, Routledge and Kegan Paul, London.

DALLY, A. (1982) *Inventing Motherhood: The Consequences of an Ideal*, Burnett, London.

DAVIDOFF, L. and HALL, C. (1987) *Family Fortunes: Men and Women of the English Middle Class 1780–1850*, Unwin Hyman, London.

DERRIDA, J. (1994) *Spectres of Marx: The State of the Debt, the Working of Mourning and the New International*, Routledge, London.

DICKEN, P., PECK, J. and TICKELL, A. (1997) Unpacking the global, in Lee, R. and Wills, J. (eds) *Geographies of Economies*, Arnold, London, pp. 158–166.

DIDCZUNEIT, V. (1997) 'Für den Sieg des Sozialismus – Ran an die Arbeit!', in Schäfer, H. (ed.) *Ungleiche Schwestern? Frauen in Ost- und Westdeutschland*, Haus der Geschichte der Bundesrepublik Deutschland, Bonn, pp. 16–23.

DÖLLING, I. (1989) Continuity and change in the media image of women in the GDR, in Gerber, M. and Woods, R. (eds) *Studies in GDR Culture and Society 8*, University Press of America, Lanham, MD, pp. 131–144.

DOWLER, L. (1998) 'And they think I'm just a nice old lady.' Women and War in Belfast, Northern Ireland, *Gender, Place and Culture*, **5**, 159–176.

DOWLING, R. and PRATT, G. (1993) Home truths: recent feminist constructions, *Urban Geography*, **14**, 464–475.

DUNFORD, M. (1998) Differential development, institutions, modes of regulation and comparative transitions to capitalism, in Pickles, J. and Smith, A. (eds) *Theorising Transition: The Political Economy of Post-Communist Transformations*, Routledge, London, pp. 76–114.

DWYER, C. (1991) *State-aided Muslim schools in Britain? Discourses of racism, nationalism and cultural pluralism*, Unpublished Masters Thesis, Syracuse University, USA.

DWYER, C. (1993) Constructions of Muslim identity and the contesting of power: the debate over Muslim schools in the UK, in Jackson, P. and Penrose, J. (eds) *Constructions of Race, Place and Nation*, UCL Press, London, pp. 143–159.

DWYER, C. (1997) *Constructions and contestations of Islam: questions of identity for young British Muslim women*, Unpublished PhD thesis, University of London, UK.

DWYER, C. (1999a) Veiled meanings: young British Muslim women and the negotiation of differences, *Gender, Place and Culture*, **6**(1), 5–26.

DWYER, C. (1999b) Contradictions of community: questions of identity for young British Muslim women, *Environment and Planning A*, **31**(1), 53–68.

DYCK, I. (1996) Mother or worker? Women's support networks, local knowledge and informal childcare strategies, in England, K. (ed.) *Who Will Mind the Baby? Geographies of Child Care and Working Mothers*, Routledge, London, pp. 123–140.

EADE, J. (1994) Identity, nation and religion: educated young Bangladeshi Muslims in London's 'East End', *International Sociology*, **9**, 377–394.

EINHORN, B. (1993) *Cinderella Goes to Market: Citizenship, Gender and Women's Movements in East Central Europe*, Verso, New York.

ELSON, D. (ed.) (1990) *Male Bias in the Development Process*, Manchester University Press, Manchester.

ELSON, D. and PEARSON, R. (1981) The subordination of women and the internationalisation of factory production, in Young, K., Wolkowitz, C. and McCullagh, R. (eds) *Of Marriage and the Market: Women's Subordination Internationally and its Lessons*, Routledge, London, pp. 18–40.

ESCOBAR, A. (1994) *Encountering Development: The Making and Unmaking of the Third World*, Princeton University Press, Princeton, NJ.

ESCOBAR, A. and ALVAREZ, S. (eds) (1992) *The Making of Social Movements in Latin America. Identity, Strategy and Democracy*, Westview Press, Boulder, CO.

EWEN, S. (1976) *Captains of Consciousness: Advertising and the Social Roots of Consumer Culture*, McGraw-Hill, New York.

FERREE, M.M. (1993) The rise and fall of 'mommy politics': feminism and unification in (East) Germany, *Feminist Studies*, **19**, 89–115.

FINCH, J. (1985) The deceit of self-help: pre-school playgroups and working class mothers, *Journal of Social Policy*, **13**, 1–20.

FINCHER, R. (1996) The state and childcare: an international review from a geographical perspective, in England, K. (ed.) *Who Will Mind the Baby? Geographies of Child Care and Working Mothers*, Routledge, London, pp. 143–166.

FISHER, J. (1993) *Out of the Shadows: Women, Resistance and Politics in South America*, Latin American Bureau, London.

FOORD, J. and GREGSON, N. (1986) Patriarchy: towards a reconceptualisation, *Antipode*, **18**, 186–211.

FOX HARDING, L. (1994) 'Parental responsibility' – a dominant theme in British child and family policy for the 1990s, *International Journal of Sociology and Social Policy*, **14**, 84–108.

FRASER, N. (1992) Rethinking the public sphere: a contribution to the critique of the actually existing democracy, in Calhoun, C. (ed.) *Habermas and the Public Sphere*, MIT Press, Cambridge, MA, pp. 109–142.

FREISTAAT SACHSEN (1996) *Wirtschaft und Arbeit in Sachsen*, Staatsministerium für Wirtschaft und Arbeit, Freistaat Sachsen, Dresden.

FREVERT, U. (1997) Weimarer Zeit und NS-Zeit, in Schäfer, H. (ed.) *Ungleiche Schwestern? Frauen in Ost- und Westdeutschland*, Haus der Geschichte der Bundesrepublik Deutschland, Bonn, pp. 8–13.

FUNK, N. and MUELLER, M. (eds) (1993) *Gender Politics and Post-communism*, Routledge, London.

GIBSON-GRAHAM, J.K. (1994) 'Stuffed if I know!': reflections on post-modern feminist social research, *Gender, Place and Culture*, **1**, 205–225.

GIBSON-GRAHAM, J.K. (1996) *The End of Capitalism (As We Knew It): A Feminist Critique of Political Economy*, Blackwell, Cambridge, MA.

GIER, J. and WALTON, J. (1987) Some problems with reconceptualising patriarchy, *Antipode*, **19**, 54–58.

GILBERT, M. (1994) The politics of location: doing feminist research at 'home', *Professional Geographer*, **46**, 90–96.

GILLESPIE, M. (1995) *Television, Ethnicity and Cultural Change*, Routledge, London.

GILROY, P. (1993) *Small Acts*, Serpent's Tail, London.

GRABHER, G. and STARK, D. (eds) (1997) *Restructuring Networks in Postsocialism*, Oxford University Press, Oxford.

GRAHAM, A. and REGULSKA, J. (1997) Expanding political space for women in Poland: an analysis of three communities, *Communist and Post-Communist Studies*, **30**, 65–82.

GREGSON, N. and FOORD, J. (1987) Patriarchy: comments on critics, *Antipode*, **19**, 371–375.

GREGSON, N. and LOWE, M. (1994) *Servicing the Middle Classes. Class, Gender and Waged Domestic Labour in Contemporary Britain*, Routledge, London.

GREGSON, N. and LOWE, M. (1995) 'Home'-making: On the spatiality of daily social reproduction in contemporary middle-class Britain, *Transactions of the Institute of British Geographers*, **20**, 224–235.

HALL, S. (1990) Cultural identity and diaspora, in Rutherford, J. (ed.) *Identity: Community, Culture, Difference*, Lawrence and Wishart, London, pp. 222–238.

HALL, S. (1992a) The question of cultural identity, in Hall, S., Held, D. and McGrew, A. (eds) *Modernity and its Futures*, Polity Press, Cambridge, pp. 273–325.

HALL, S. (1992b) New ethnicities, in Donald, J. and Rattansi, A. (eds) *'Race', Culture and Difference*, Sage, London, pp. 252–259. Originally published in Mercer, K. (ed.) (1988) *Black Film, British Cinema*, Institute of Contemporary Arts, London, pp. 27–31.

HALL, S. (1995) New cultures for old, in Massey, D. and Jess, P. (eds) *A Place in the World?*, Oxford University Press, Oxford, pp. 175–213.

HAMPELE, A. (1993) The organised women's movement in the collapse of the GDR: the Independent Women's Association (UFV), in Funk, N. and Mueller, M. (eds) *Gender Politics and Post-communism*, Routledge, London, pp. 194–200.

HANSON, S. and PRATT, G. (1988) Reconceptualising the links between home and work in urban geography, *Economic Geography*, 43, 299–321.

HARAWAY, D. (1991) *Simians, Cyborgs and Women: The Reinvention of Nature*, Free Association Books, London.

HARAWAY, D. (1997) *Modest_Witness@Second_Millennium.FemaleMan_Meets_ OncoMouse™: Feminism and Technoscience*, Routledge, London.

HARDING, S. (1987) Introduction: Is there a feminist method?, in Harding, S. (ed.) *Feminism and Methodology*, Indiana University Press, Bloomington, IN, pp. 1–14.

HARVEY, D. (1989) *The Condition of Postmodernity*, Blackwell, Oxford.

HARVEY, D. (1993) Class relations, social justice and the politics of difference, in Keith, M. and Pile, S. (eds) *Place and the Politics of Identity*, Routledge, London, pp. 41–66.

HAW, K. (1996) Exploring the educational experiences of Muslim girls: tales told to tourists – should the white researcher stay at home?, *British Educational Research Journal*, 22, 319–330.

HIGGINS, A. (1998) Uncle Sam disturbs Asia's dreams, *The Observer*, 18 January, p. 10.

HILDEBRANDT, R. (1996) Zu Hause fällt uns die Decke auf den Kopf, in Wolf, C. (ed.) *Frauen und Marktwert*, Wichern, Berlin, pp. 38–41.

HIRST, P. and THOMPSON, G. (1996) *Globalisation in Question*, Polity, Cambridge.

HÖHMANN, H.-H. and SEIDENSTREICHER, G. (1980) Partizipation im System der administrativen Planwirtschaft von UdSSR und DDR, in Höhmann, H.H. (ed.) *Partizipation und Wirtschaftsplanung in Osteuropa und der PR China*, Kohlhammer, Stuttgart, pp. 9–52.

HOLLOWAY, S.L. (1996) *Space, place and geographies of childcare*, Unpublished PhD thesis, University of Sheffield, UK.

HOLLOWAY, S.L. (1998a) Geographies of justice: preschool-childcare provision and the conceptualisation of social justice, *Environment and Planning C: Government and Policy*, 16, 85–104.

HOLLOWAY, S.L. (1998b) Local childcare cultures: moral geographies of mothering and the social organisation of pre-school education, *Gender, Place and Culture*, 5, 29–53.

HOLLOWAY, S.L. (1998c) 'She lets me go out once a week': mothers' strategies for obtaining 'personal' time and space, *Area*, 30, 321–330.

HOLLOWAY, S.L. (1999) Mother and worker?: the negotiation of motherhood and paid employment in two urban neighbourhoods, *Urban Geography*, 20.

HOLLOWAY, S.L., VALENTINE, G. and BINGHAM, N. (1998) Children's social networks, 'virtual communities' and on-line spaces, paper presented at the Association of American Geographers Annual Meeting, Boston, MA, 25–29 March.

HOOKS, b. (1991) *Yearning: Race, Gender and Cultural Politics*, Southend Press, Boston, MA.

HOOKS, b. (1992) *Black Looks: Race and Representation*, Turnaround, London.

HÖRSCHELMANN, K. (1997) Watching the East: constructions of 'otherness' in TV representations of East Germany, *Applied Geography*, 17, 385–396.

HORSMAN, M. and MARSHALL, A. (1994) *After the Nation-state: Citizens, Tribalism and the New World Disorder*, Harper Collins, London.

JACKSON, P. (1987) *Race and Racism*, Allen and Unwin, London.

JACKSON, P. (1991) The cultural politics of masculinity: towards a social geography, *Transactions of the Institute of British Geographers*, 16, 199–213.

JACKSON, P. (1993) Towards a cultural politics of consumption, in Bird, J., Curtis, B., Putnam, T., Robertson, G. and Tickner, L. (eds) *Mapping the Futures: Local Cultures, Global Change*, Routledge, London, pp. 207–228.

JACOBSON, J. (1997) Religion and ethnicity: dual and alternative sources of identity among young British Pakistanis, *Ethnic and Racial Studies*, 20, 238–256.

JAGGAR, A.M. (1984) Human biology in feminist theory: sexual equality reconsidered, in Gould, C. (ed.) *Beyond Domination: New Perspectives on Women and Philosophy*, Rowman and Allanheld, Totowa, NJ, pp. 21–42.

JELIN, E. (ed.) (1990) *Women and Social Change in Latin America*, Zed Press, London.

JESSOP, B. (1993) Towards a Schumpetarian workfare state? Preliminary remarks on post-Fordist political economy, *Studies in Political Economy*, 40, 7–39.

JOHNSON, L. (1987) (Un)realist perspectives: patriarchy and feminist challenges in geography, *Antipode*, 19, 210–215.

JOHNSON, N. (1995) Cast in stone: monuments, geography and nationalism, *Environment and Planning D: Society and Space*, 13, 51–65.

JOHNSTON, L. (1998) Reading the sexed bodies and spaces of gyms, in Nast, H. and Pile, S. (eds) *Places Through the Body*, Routledge, London, pp. 244–262.

JOHNSTON, L. and VALENTINE, G. (1995) Wherever I lay my girlfriend, that's my home: the performance and surveillance of lesbian identities in domestic environments, in Bell, D. and Valentine, G. (eds) *Mapping Desire: Geographies of Sexualities*, Routledge, London, pp. 99–113.

JOLY, D. (1995) *Britannia's Crescent: Making a Place for Muslims in British Society*, Avebury, Aldershot.

JONES, J.P., NAST, H. and ROBERTS, S. (1997) *Thresholds in Feminist Geography: Difference, Methodology, Representation*, Rowman and Littlefield, New York.

KABEER, N. (1994) *Reversed Realities: Gender Hierarchies in Development Thought*, Verso, London.

KAHLAU, C. (ed.) (1990) *Aufbruch! Frauenbewegung in der DDR: Dokumentation*, Frauenoffensive, München.

KARL, M. (1995) *Women and Empowerment: Participation and Decision Making*, Zed Press, London.

KASSAM, N. (1997) *Telling It Like It Is: Young Asian Women Talk*, The Women's Press, London.

KATZ, C. (1994) Playing the field: questions of fieldwork in geography, *Professional Geographer*, 46, 67–72.

KATZ, C. and MONK, J. (eds) (1993) *Full Circles: Geographies of Women over the Life Course*, Routledge, London.

KEARNS, A. (1995) Active citizenship and local governance: political and geographical dimensions, *Political Geography*, 14, 155–175.

KING, R. (1995) Migrations, globalisation and place, in Massey, D. and Jess, P. (eds) *A Place in the World?*, Oxford University Press, Oxford, pp. 5–44.

KITCHIN, R. (1998) *Cyberspace*, John Wiley, Chichester.

KNOPP, L. and LAURIA, M. (1987) Gender relations as a particular form of social relations, *Antipode*, **19**, 48–53.

KNOTT, K. and KHOKHER, S. (1993) Religion and ethnic identity among young Muslim women in Bradford, *New Community*, **19**, 593–610.

KOBAYASHI, A. (1994) Coloring the field: gender, 'race' and the politics of fieldwork, *Professional Geographer*, **46**, 73–80.

KOBAYASHI, A. and PEAKE, L. (1994) Unnatural discourse. 'Race' and gender in geography, *Gender, Place and Culture*, **1**, 225–244.

KOFMAN, E. (1995) Citizenship for some but not for others: spaces of citizenship in contemporary Europe, *Political Geography*, **14**, 121–137.

KOLINSKY, E. (1993) *Women in Contemporary Germany: Life, Work and Politics*, Berg, Providence, Oxford.

KRECKEL, R. (1995) Makrosoziologische Überlegungen zum Kampf um Normal- und Teilzeitarbeit im Geschlechterverhältnis, *Berliner Journal für Soziologie*, **4**, 489–495.

LAPP, P.J. (1988) *Die Blockparteien im politischen System der DDR*, Verlag Ernst Knoth, Melle.

LAURIE, N. (1995) *Negotiating gender: women and emergency employment in Peru*, Unpublished PhD thesis, University of London, UK.

LAURIE, N. (1997a) Negotiating femininity: women and representation in emergency employment in Peru, *Gender, Place and Culture*, **4**, 235–251.

LAURIE, N. (1997b) From work to welfare: the response of the Peruvian state to the feminization of emergency work, *Political Geography*, **16**, 691–714.

LAURIE, N. (1999) Negotiating femininities in the 'provinces'. Women and emergency employment in Peru, *Environment and Planning A*, **31**, 229–250.

LAURIE, N. and MARVIN, S. (forthcoming) Globalisation, neo-liberalism and negotiated development in the Andes. Water projects and regional identity in Cochabamba, Bolivia, *Environment and Planning 'A'*.

LAURIE, N. and SMITH, F.M. (1998) More than hinnies' geographies? Encounters between feminist and political geography, paper presented at the Annual Conference of the Royal Geographical Society with the Institute of British Geographers, Guildford, 4–7 January.

LAWS, S. (1994) Un-valued families, *Trouble and Strife*, **28**, 5–11.

LAZREG, M. (1988) Feminism and difference: the perils of writing as a woman on women in Algeria, *Feminist Studies*, **14**, 81–107.

LEE, R. and WILLS, J. (eds) (1997) *Geographies of Economies*, Arnold, London.

LEES, S. (1993) *Sugar and Spice: Sexuality and Adolescent Girls*, Penguin, London.

LEMKE, C. (1991) *Die Ursachen des Umbruchs 1989: politische Sozialisation in der ehemaligen DDR*, Westdeutscher Verlag, Opladen.

LESLIE, D.A. (1993) Femininity, post-Fordism and the 'New Traditionalism', *Environment and Planning D: Society and Space*, **11**, 689–708.

LETHERBY, G. (1994) Mother or not, mother or what?: problems of definition and identity, *Women's Studies International Forum*, **17**, 525–532.

LEWIS, P. (1994) *Islamic Britain*, I. B. Taurus, London.

LEYSHON, A. (1995) Annihilating space? The speed up of communications, in Allen, T. and Hamnett, C. (eds) *A Shrinking World? Global Unevenness and Inequality*, Oxford University Press, Oxford, pp. 12–46.

LEYSHON, A. and THRIFT, N. (1996) *Money/space: Geographies of Monetary Transformation*, Routledge, London.

LI, F.L., FINDLAY, A.M. and JONES, H. (1998) A cultural economy perspective on service sector migration in the global city: the case of Hong Kong, *International Migration*, **36**, 131–157.

LIGHT, A. (1991) *Forever England: Femininity, Literature and Conservatism Between the Wars*, Routledge, London.

LIGHT, J.S. (1995) The digital landscape: a new space for women?, *Gender, Place and Culture*, **2**, 133–146.

LIM, L. (1990) Women's work in export factories: the politics of a cause, in Tinker, I. (ed.) *Persistent Inequalities: Women and World Development*, Oxford University Press, New York, pp. 101–122.

LONGHURST, R. (1997) (Dis)embodied geographies, *Progress in Human Geography*, **21**, 486–501.

LONGHURST, R. (1998) (Re)presenting shopping centres and bodies: questions of pregnancy, in Ainley, R. (ed.) *New Frontiers of Space, Bodies and Gender*, Routledge, London, pp. 20–34.

LUKE, T. (1997) Localized spaces, globalized places: virtual community and geo-economics in the Asia-Pacific, in Berger M. and Borer, D. (eds) *The Rise of East Asia: Critical Visions of the Pacific Century*, Routledge, London, pp. 241–259.

MACKENZIE, S. (1984) Editorial introduction, *Antipode*, **16**, 15–17.

MACLEOD, A.E. (1992) Hegemonic relations and gender resistance: the new veiling as accommodating protest in Cairo, *Signs*, **17**, 533–554.

MADGE, C. (1993) Boundary disputes: comments on Sidaway (1992), *Area*, **25**, 294–299.

MALECK-LEWY, E. (1995) Between self-determination and state supervision: women and abortion law in post-unification Germany, *Social Politics*, **2**, 62–75.

MANI, L. (1992) Cultural theory, colonial texts, in Grossberg, L., Neilson, C. and Treichler, P. (eds) *Cultural Studies*, Routledge, New York, pp. 392–408.

MARCHARD, M. (1994) Gender and new regionalism in Latin America: inclusion/exclusion, *Third World Quarterly*, **15**, 63–76.

MARK-LAWSON, J., SAVAGE, M. and WARDE, A. (1985) Gender and local politics: struggles over welfare policies, 1918–1939, in The Lancaster Regionalism Group, *Localities, Class and Gender*, Pion, London, pp. 195–215.

MARSHALL, H. (1991) The social construction of motherhood: an analysis of child-care and parenting manuals, in Phoenix, A., Woollett, A. and Lloyd, E. (eds) *Motherhood: Meanings, Practices and Ideologies*, Sage, London, pp. 66–85.

MARSHALL, J.N. and RICHARDSON, R. (1996) The impact of 'telemediated' services on corporate structures, *Environment and Planning A*, **28**, 1843–1858.

MARTIN, J. and ROBERTS, C. (1984) *Women and Employment: A Lifetime Perspective*, HMSO, London.

MARTIN, P. (1997) Saline politics: local participation and neoliberalism in Australian rural environments, *Space and Polity*, **1**, 115–133.

MARTIN, S. (1992) *Refugee Women*, Zed Press, London.

MARVIN, S. and GRAHAM, S. (1994) Privatisation of utilities: the implications for cities in the United Kingdom, *Journal of Urban Technology*, **2**, 47–66.

MASSEY, D. (1994) *Space, Place and Gender*, Polity Press, Cambridge.

MASSEY, D. (1995) Masculinity, dualisms and high technology, *Transactions of the Institute of British Geographers*, **20**, 487–499.

MASSEY, D. (1996) Politicising space and place, *Scottish Geographical Magazine*, **112**, 117–123.

MASSEY, D. and JESS, P. (1995) Places and cultures in an uneven world, in Massey, D. and Jess, P. (eds) *A Place in the World?*, Oxford University Press, Oxford, pp. 45–77.

MAVILA LEÓN, R. (1992) Presente y futuro del las mujeres de la guerra, *QueHacer*, 44–49.

MAY, J. (1996) A little taste of something more exotic, *Geography*, 81, 57–64.

MAZUMDAR, S. (1995) Women on the march: right-wing mobilization in contemporary India, *Feminist Review*, 49, 1–28.

McDOWELL, L. (1983) Towards an understanding of the gender division of urban space, *Environment and Planning D: Society and Space*, 1, 59–72.

McDOWELL, L. (1986) Beyond patriarchy: a class-based explanation of women's subordination, *Antipode*, 18, 311–321.

McDOWELL, L. (1991) Life without father and Ford: the new gender order of post-Fordism, *Transactions of the Institute of British Geographers*, 16, 400–419.

McDOWELL, L. (1992) Doing gender: feminism, feminists and research methods in human geography, *Transactions of the Institute of British Geographers*, 17, 399–416.

McDOWELL, L. (1993) Space, place and gender relations: Part II. Identity, difference, feminist geometries and geographies, *Progress in Human Geography*, 17, 305–318.

McDOWELL, L. (1995) Body work: heterosexual gender performances in city workplaces, in Bell, D. and Valentine, G. (eds) *Mapping Desire*, Routledge, London, pp. 75–95.

McDOWELL, L. and COURT, G. (1994a) Missing subjects: gender, sexuality and power in merchant banking, *Economic Geography*, 70, 229–251.

McDOWELL, L. and COURT, G. (1994b) Performing work: bodily representations in merchant banks, *Environment and Planning D: Society and Space*, 12, 727–750.

McDOWELL, L. and MASSEY, D. (1984) A woman's place?, in Massey, D. and Allen, J. (eds) *Geography Matters: A Reader*, Cambridge University Press, Cambridge, pp. 128–147.

McDOWELL, L. and SHARP, J. (eds) (1997) *Space, Gender, Knowledge*, Arnold, London.

McKINSEY AND COMPANY (1997) *The Future of Information Technology in UK Schools*, McKinsey and Company, London.

McLAFFERTY, S.L. (1995) Counting for women, *Professional Geographer*, 47, 436–442.

McLOUGHLIN, S. (1996) In the name of the Umma: globalization, 'race' relations and Muslim identity politics in Bradford, in Shadid, W.A.R. and van Koningsveld, P.S. (eds) *Political Participation and Identities of Muslims in Non-Muslim States*, Kok Pharos, Kampen, pp. 206–228.

McRAE, S. (1991) Occupational change over childbirth: evidence from a national survey, *Sociology*, 25, 589–605.

McROBBIE, A. (1991) *Feminism and Youth Culture*, Macmillan, London.

McROBBIE, A. (1993) Shut up and dance: youth culture and changing modes of femininity, *Cultural Studies*, 7, 406–426.

McROBBIE, A. (1994) Folk devils fight back, *New Left Review*, 203, 107–116.

McROBBIE, A. (1996) Different, youthful subjectivities, in Chambers, I. and Curtis, L. (eds) *The Postcolonial Question*, Routledge, London, pp. 30–46.

McROBBIE, A. and NAVA, M. (eds) (1984) *Gender and Generation*, Macmillan, London.

MELTZER, H. (1994) *Day Care Services for Children: A Survey Carried Out on Behalf of the Department of Health in 1990*, HMSO, London.

MILLAR, J. (1994) State, family and personal responsibility: the changing balance for lone mothers in the United Kingdom, *Feminist Review*, 48, 24–39.

MIRZA, K. (1989) *The Silent Cry: Second Generation Bradford Muslim Women Speak*, Centre for the Study of Islam and Christian–Muslim Relations, Selly Oak Colleges, Birmingham, Research Paper No. 43.

MODOOD, T. (1988) 'Black', racial equality and Asian identity, *New Community*, 14, 397–404.

MODOOD, T. (1992) British Muslims and the Rushdie Affair, in Donald, J. and Rattansi, A. (eds) *'Race', Culture and Difference*, Sage, London, pp. 260–277.

MOHAMMAD, R. (1996) Insider/outsider research and the politics of representation, paper presented at the Feminist Methodologies Conference, Nottingham Trent University, 9 March.

MOHANTY, C. (1991) Under Western eyes: feminist scholarship and colonial discourse, in Mohanty, C., Russo, A. and Torres, L. (eds) *Third World Women and the Politics of Feminism*, Indiana University Press, Bloomington, IN, pp. 51–80.

MONK, J. and HANSON, S. (1982) On not excluding half of the human in human geography, *Professional Geographer*, 34, 11–23.

MORAGA, C. and ANZALDUA, G. (eds) (1991) *This Bridge Called My Back: Writings by Radical Women of Colour*, Kitchen Table, New York (2nd edition).

MORLEY, D. and ROBINS, K. (1995) *Spaces of Identity: Global Media, Electronic Landscapes and Cultural Boundaries*, Routledge, London.

MOSER, C. (1981) Surviving in the suburbios, *Bulletin of the Institute of Development Studies*, 12, 19–29.

MOSER, C. (1993) Adjustment from below: low-income women, time and the triple role in Guayaquil, Ecuador, in Radcliffe, S. and Westwood, S. (eds) *'Viva' Women and Popular Protest in Latin America*, Routledge, London, pp. 173–196.

MOSS, P. (1993) Focus: feminism as method, *Canadian Geographer*, 37, 48–49.

MOSS, P. (ed.) (1995) *Father Figures: Fathers in the Families of the 1990s*, HMSO, Edinburgh.

MOTOROLA (1997) *The British and Technology*, Motorola, Slough.

MUJER Y AJUSTE (1996) *Ajuste Estructural Debate y Propuestas. El Ajuste Estructural en el Perú; una Mirada Desde las Mujeres*, Ediciones Mujeres y Ajuste, Lima.

NAST, H.J. (1994) Opening remarks on 'women in the field', *Professional Geographer*, 46, 54–66.

NÈVE, D. DE (1997) Politische Partizipation von Frauen in Rumänien nach 1989, *Osteuropa*, 2, 160–172.

NEWSON, J. and NEWSON, E. (1974) Cultural aspects of childrearing in the English-speaking world, in Richards, M.P.M. (ed.) *The Integration of a Child into a Social World*, Cambridge University Press, Cambridge.

NICKEL, H.M. (1990) Geschlechtertrennung durch Arbeitsteilung: Berufs- und Familienarbeit in der DDR, *Feministische Studien*, 8, 10–19.

NICKEL, H.M. (1992) Frauenarbeit in den neuen Bundesländern: Rück- und Ausblick, *Berliner Journal für Soziologie*, 2, 39–48.

NICKEL, H.M. (1993a) Mitgestalterinnen des Sozialismus: Frauenarbeit in der DDR, in Helwig, G. and Nickel, H.M. (eds) *Frauen in Deutschland 1945–1992*, Akademie-Verlag, Berlin.

NICKEL, H.M. (1993b) Women in the German Democratic Republic and in the New Federal States: Looking Backward and Forward (Five Theses), in Funk, N. and Mueller, M. (eds) *Gender Politics and Post-Communism*, Routledge, London, pp. 38–150.

OHMAE, K. (1990) *The Borderless World: Power and Strategy in an Interlinked World*, Harper Collins, New York.

OHMAE, K. (1995) *The End of the Nation State: The Rise of Regional Economies*, Harper Collins, London.

ONS (Office for National Statistics) (1998) *Social Trends 28*, TSO, London.

OPCS (Office of Population Censuses and Surveys) (1993) *General Household Survey, 1991*, HMSO, London.

OSTERLAND, M. (1994) Coping with democracy: the re-institution of self-government in Eastern Germany, *European Urban and Regional Research*, 1, 5–18.

Ó TUATHAIL, G. (1998) Political geography III: dealing with deterritorialization, *Progress in Human Geography*, 22, 81–93.

Ó TUATHAIL, G. and LUKE, T. (1994) Present at the disintegration: deterritorialisation and reterritorialisation in the new world order, *Annals of the Association of American Geographers*, 84, 381–398.

OWEN, C. and MOSS, P. (1989) Patterns of pre-school provision in English local authorities, *Journal of Education Policy*, 4, 309–328.

OWEN, M. (1996) *A World of Widows*, Zed Press, London.

PALMER, D. (ed.) (1992) *The Shining Path of Peru*, Hurst and Company, London.

PAREDES, P. and TELLO, G. (eds) (1988) *Pobreza Urbana y Trabajo Feminino*, ADEC and ATC, Lima.

PARMAR, P. (1984) Hateful contraries: media images of Asian women, *Ten*, 8, 71–78.

PEARSON, R. (1986) Latin American women and the New International Division of Labour: a reassessment, *Bulletin of Latin American Research*, 5, 67–79.

PEARSON, R. and MITTER, S. (1993) Employment and working conditions of low-skilled information-processing workers in less developed countries, *International Labour Review*, 132, 49–64.

PECK, J. and TICKELL, A. (1994) Jungle law breaks out: neoliberalism and global–local disorder, *Area*, 26, 317–326.

PENROSE, V. (1993) The political participation of GDR women during the *Wende*, in Gerber, M. and Woods, R. (eds) *Studies in GDR Culture and Society 11/12*, University Press of America, Lanham, MD, pp. 37–52.

PETTMAN, J. (1996) An international economy of sex?, in Kofman, E. and Youngs, G. (eds) *Globalisation: Theory and Practice*, Routledge, London, pp. 191–208.

PHILO, C. (1991) De-limiting human geography: new social and cultural perspectives, in Philo, C. (compiler) *New Words, New Worlds: Reconceptualising Social and Cultural Geography*, Cambrian Printers, Aberystwyth, pp. 14–27.

PHOENIX, A. (1991) *Young Mothers?*, Polity Press, Cambridge.

PHOENIX, A. (1994) Practising feminist research: the intersection of gender and 'race' in the research process, in Maynard, M. and Purvis, J. (eds) *Researching Womens' Lives from a Feminist Perspective*, Taylor and Francis, London, pp. 49–71.

PHOENIX, A. and WOOLLETT, A. (1991a) Introduction, in Phoenix, A., Woollett, A. and Lloyd, E. (eds) *Motherhood: Meanings, Practices and Ideologies*, Sage, London, pp. 1–12.

PHOENIX, A. and WOOLLETT, A. (1991b) Motherhood: social construction, politics and psychology, in Phoenix, A., Woollett, A. and Lloyd, E. (eds) *Motherhood: Meanings, Practices and Ideologies*, Sage, London, pp. 13–27.

PILE, S. (1994) Masculinism, the use of dualistic epistemologies and third spaces, *Antipode*, 26, 255–277.

PILE, S. and THRIFT, N. (eds) (1995) *Mapping the Subject: Geographies of Cultural Transformation*, Routledge, London.

PILKINGTON, H. (1997) 'For the sake of the children': gender and migration in the former Soviet Union, in Buckley, M. (ed.) *Post-Soviet Women: From the Baltic to Central Asia*, Cambridge University Press, Cambridge, pp. 119–140.

POOLE, D. and RENIQUE, G. (1992) *Peru: Time of Fear*, Latin American Bureau, London.

PRATT, G. and HANSON, S. (1995) *Gender, Work and Space*, Routledge, London.

PRATT, G. in collaboration with the Philippine Women Centre, Vancouver, Canada (1998) Inscribing domestic work on Filipina bodies, in Nast, H. and Pile, S. (eds) *Places Through the Body*, Routledge, London, pp. 283–304.

PRINGLE, R. (1989) *Secretaries Talk: Sexuality, Power and Work*, Verso, London.

Professional Geographer (1994) Women in the field: critical feminist methodologies and theoretical perspectives, Special Issue, Nast, H.J. (ed.), **46**, 55–102.

Professional Geographer (1995) Should women count? The role of quantitative methodology in feminist geographical research, Special Issue, Hodge, D.C. (ed.), **47**, 426–466.

QUACK, S. and MAIER, F. (1994) From state-socialism to market economy – women's employment in East Germany, *Environment and Planning A*, **26**, 1257–1276.

RADCLIFFE, S. (1986) Gender relations, peasant livelihood strategies and migration: a case study from Cuzco, Peru, *Bulletin of Latin American Research*, **5**, 29–47.

RADCLIFFE, S. (1993) 'People have to rise up – like the great women fighters': the state and peasant women in Peru, in Radcliffe, S. and Westwood, S. (eds) *'Viva' Women and Popular Protest in Latin America*, Routledge, London, pp. 197–218.

RADCLIFFE, S. (1994) (Representing) post-colonial women: authenticity, difference and feminism, *Area*, **26**, 25–32.

RADCLIFFE, S. (1998) Frontiers and popular nationhood: geographies of identity in the 1995 Ecuador–Peru border dispute, *Political Geography*, **17**, 273–293.

RADCLIFFE, S. and WESTWOOD, S. (eds) (1993) *'Viva' Women and Popular Protest in Latin America*, Routledge, London.

RADCLIFFE, S. and WESTWOOD, S. (1996) *Remaking the Nation: Place Identity and Politics in Latin America*, Routledge, London.

RAI, A. (1995) India on-line: electronic bulletin boards and the construction of a diasporic Hindu identity, *Diaspora*, **4**, 31–58.

RAMAZANOGLU, C. (1989) *Feminism and the Contradictions of Oppression*, Routledge, London.

REDCLIFT, N. and MINGIONE, E. (eds) (1985) *Beyond Employment. Household, Gender and Subsistence*, Basil Blackwell, Oxford.

REGULSKA, J. (1998) 'The political' and its meaning for women: transition politics in Poland, in Pickles, J. and Smith, A. (eds) *Theorising Transition: The Political Economy of Post-communist Transformations*, Routledge, London, pp. 309–329.

REIMANN, B. (1997) The transition from people's property to private property, *Applied Geography*, **17**, 301–314.

RENNE, T. (ed.) (1997) *Ana's Land: Sisterhood in Eastern Europe*, Westview Press, Boulder, CO.

RHODES, P. (1994) Race-of-interviewer effects: a brief comment, *Sociology*, **28**, 547–558.

RIBBENS, J. (1989) Interviewing: an 'unnatural situation', *Women's Studies International Forum*, **12**, 579–592.

RICHARDSON, D. (1993) *Women, Motherhood and Childrearing*, Macmillan, London.

RIEMER, J. (1992) Reproduction and reunification: the politics of abortion in united Germany, in Huelshoff, M. and Markovits, A. (eds) *The New Germany in the New Europe*, University of Michigan Press, Ann Arbor, MI.

ROSA, K. (1994) The conditions and organisational activities of women in Free Trade Zones; Malaysia, Philippines and Sri Lanka, 1970–1990, in Rowbotham, S. and Mitter, S. (eds) *Dignity and Daily Bread. New Forms of Economic Organising Among Poor Women in the Third World and the First*, Routledge, London, pp. 73–100.

ROSA-LUXEMBURG-VEREIN E.V. (1992) *Frauen in Sachsen: zwischen Betroffenheit und Hoffnung*, Rosa-Luxemburg-Verein e.V., Leipzig.

ROSE, D. (1993) On feminism, method and methods in human geography: an idiosyncratic overview, *Canadian Geographer*, 37, 57–61.

ROSE, G. (1993) *Feminism and Geography*, Polity, Cambridge.

ROSE, G. (1994) The cultural politics of place: local representation and oppositional discourses in two films, *Transactions of the Institute of British Geographers*, 19, 46–60.

ROSE, G. (1997) Situating knowledges: positionality, reflexivities and other tactics, *Progress in Human Geography*, 21, 305–320.

ROWLANDS, J. (1997) *Questioning Empowerment*, Oxfam, Oxford.

RUGGIE, J.G. (1993) Territoriality and beyond: problematizing modernity in international relations, *International Organization*, 47, 139–174.

RUIZ BRAVO, P. (1998) New challenges to the women's movement in Peru, paper presented at Department of Geography Seminar, Newcastle University, 28 April.

RUTHERFORD, J. (1990) A place called home: identity and the cultural politics of difference, in Rutherford, J. (ed.) *Identity: Community, Culture, Difference*, Lawrence and Wishart, London, pp. 9–27.

RUTVEN, M. (1990) *A Satanic Affair: Salman Rushdie and the Wrath of Islam*, Hogarth Press, London.

SAMAD, Y. (1993) Imagining a British Muslim identification, paper presented at 'Muslims in Europe: Generation to Generation' Conference, St Catherine's College, Oxford, 5–7 April.

SASSEN, S. (1991) *The Global City*, Princeton University Press, Princeton, NJ.

SCHÄFER, H. (ed.) (1997) *Ungleiche Schwestern? Frauen in Ost- und Westdeutschland*, Haus der Geschichte der Bundesrepublik Deutschland, Bonn.

SCHENK, C. (1993) Lesbians and their emancipation in the former German Democratic Republic: past and future, in Funk, N. and Mueller, M. (eds) *Gender Politics and Post-communism*, Routledge, London, pp. 160–167.

SCHENK, S. (1995) Neu- oder Restrukturierung des Geschlechterverhältnisses in Ostdeutschland, *Berliner Journal für Soziologie*, 4, 475–488.

SCHMUDE, J. (1996) Contrasting developments in female labour force participation in East and West Germany since 1945, in García-Ramon, M.D. and Monk, J. (eds) *Women of the European Union*, Routledge, London, pp. 156–185.

SCOTT, A. (1994) *Divisions and Solidarities: Gender, Class and Employment in Latin America*, Routledge, London.

SCOTT, J. (1998) *Seeing Like a State*, Yale University Press, Cambridge, MA.

SCOTT, J., BRAUN, M. and ALWIN, D. (1993) The family way, in Jowell, R., Brook, L. and Dowds, L. with Ahrendt, D. (eds) *International Social Attitudes: the 10th BSA Report*, Dartmouth, Aldershot.

SCOTT, J.W., KAPLAN, C. and KEATES, D. (eds) (1997) *Transitions, Environments, Translations: Feminisms in International Politics*, Routledge, New York.

SHARMA, S., HUTNYK, J. and SHARMA, A. (1996) *Dis-orienting Rhythms: the Politics of the New Asian Dance Music*, Zed Books, London.

SHARP, J. (1996) Gendering nationhood: a feminist engagement with national identity, in Duncan, N. (ed.) *Bodyspace: Destabilizing Geographies of Gender and Sexuality*, Routledge, London, pp. 97–108.

SHARPE, S. (1994) *'Just Like a Girl'*, Penguin, London.

SHAW, A. (1988) *A Pakistani Community in Britain*, Basil Blackwell, Oxford.

SHEPHERD, M. (1998) The (re)negotiation of masculinity in Sheffield, UK, in Unwin, T. (ed.) *A European Geography*, Longman, Harlow, pp. 286–288.

SIDAWAY, J. (1992) In other worlds: on the politics of research by 'First World' geographers in the 'Third World', *Area*, 24, 403–408.

SIDAWAY, J. and POWER, M. (1998) 'Sex and violence on the wild frontiers': the aftermath of state socialism in the periphery', in Pickles, J. and Smith, A. (eds) *Theorising Transition: The Political Economy of Post-communist Transformations*, Routledge, London, pp. 408–427.

SIKKINK, K. and KECK, M. (1998) *Activism Beyond Borders: Advocacy Networks in International Politics*, Cornell University Press, Ithaca, NY.

SKEGGS, B. (1997) *Formations of Class and Gender*, Sage, London.

SKELTON, T. and VALENTINE, G. (eds) (1997) *Cool Places: Geographies of Youth Cultures*, Routledge, London.

SMITH, A. and PICKLES, J. (1998) Introduction: theorising transition and the political economy of transformation, in Pickles, J. and Smith, A. (eds) *Theorising Transition: the Political Economy of Post-communist Transformations*, Routledge, London, pp. 1–24.

SMITH, A. and SWAIN, A. (1997) Geographies of transformation: approaching regional economic restructuring in central and eastern Europe', in Turnock, D. (ed.) *Frameworks for Understanding Post-socialist Processes*, Department of Geography, Leicester University, Occasional Paper No. 36, pp. 32–37.

SMITH, A. and SWAIN, A. (1998) Regulating and institutionalising capitalisms: the micro-foundations of transformation in Eastern and Central Europe, in Pickles, J. and Smith, A. (eds) *Theorising Transition: The Political Economy of Post-communist Transformations*, Routledge, London, pp. 25–53.

SMITH, F.M. (1996a) Problematising language: limitations and possibilities in 'foreign language' research, *Area*, 28, 160–166.

SMITH, F.M. (1996b) *Contested geographies of German reunification: neighbourhood action in Leipzig, 1989–1993*, unpublished PhD thesis, University of Glasgow, UK.

SMITH, F.M. (1996c) Housing tenures in transformation: questioning geographies of ownership in eastern Germany, *Scottish Geographical Magazine*, 112, 3–10.

SMITH, F.M. (1997) Contested geographical imaginings of reunification: a case study of urban change in Leipzig, *Applied Geography*, 17, 355–369.

SMITH, F.M. (1999a) Discourses of citizenship in transition: scale, politics and urban renewal, *Urban Studies*, 36, 167–187.

SMITH, F.M. (1999b) Post-socialism, in McDowell, L. and Sharp, J. (eds) *A Feminist Glossary of Human Geography*, Arnold, London, pp. 213–214.

SMITH, F.M. (1999c) The neighbourhood as site for contesting German reunification, in Sharp, J., Routledge, P., Philo, C. and Paddison, R. (eds) *Entanglements of Power: Geographies of Domination and Resistance*, Routledge, London.

SMITH, G. (1993) Ends, geopolitics and transitions, in Johnston, R.J. (ed.) *The Challenge for Geography*, Blackwell, Oxford, pp. 76–99.

SMITH, N. (1993) Homeless/global: scaling places, in Bird, J., Curtis, B., Putnam, T., Robertson, G. and Tickner, L. (eds) *Mapping the Futures: Local Cultures, Global Change*, Routledge, London, pp. 87–119.

SMITH, N. and KATZ, C. (1993) Grounding metaphor: towards a spatialized politics, in Keith, M. and Pile, S. (eds) *Place and the Politics of Identity*, Routledge, New York, pp. 67–83.

SPAIN, D. (1992) *Gendered Spaces*, University of North Carolina Press, Chapel Hill NC.

SPARKE, M. (1996) Displacing the field in fieldwork: masculinity, metaphor and space, in Duncan, N. (ed.) *BodySpace*, Routledge, London, pp. 212–233.

SPICE GIRLS (1997) *Girlpower*, Zone, London.

STAEHELI, L. and LAWSON, V, (1994) A discussion of 'women in the field': the politics of feminist fieldwork, *Professional Geographer*, **46**, 96–102.

STEPHEN, L. (1997) *Women and Social Movements in Latin America: Power from Below*, University of Texas Press, Austin, TX.

STEVENS, E. (1973) *Marianismo*: the other face of *machismo* in Latin America, in Pescatello, A. (ed.) *Female and Male in Latin America*, University of Pittsburg Press, Pittsburg, PA.

STICHTER, S. and PARPART, J. (1990) *Women Employment and the Family in the International Division of Labour*, Macmillan, Basingstoke.

STUDZINSKI, K. (1994) *Lesbians Talk Left, Politics*, Scarlet Press, London.

SUMMERVILLE, P.I. (1993) 'What the hell are you doing here?' or Where do I fit into the IBG?, *Women and Geography Study Group Newsletter*, **2**, 4–7.

TANSKI, J. (1994) The impact of crisis, stabilisation and structural adjustment on women in Lima, Peru, *World Development*, **22**, 1627–1642.

TAYLOR, I., EVANS, K. and FRASER, P. (1996) *A Tale of Two Cities: Global Change, Local Feeling and Everyday Life in the North of England. A Study in Manchester and Sheffield*, Routledge, London.

THRIFT, N. and LEYSHON, A. (1994) 'A phantom state?' The detraditionalisation of money, the international financial system and international finance centres, *Political Geography*, **13**, 299–327.

TÓTH, O. (1993) No envy, no pity, in Funk, N. and Mueller, M. (eds) *Gender Politics and Post-communism*, Routledge, London, pp. 213–223.

TUC (Trades Union Congress) (1997) *Flower Producers are not 'Smelling of Roses'*, Trades Union Congress, London.

UNCTAD (1997) *Trade and Development Report. Globalization, Distribution and Growth*, United Nations, Geneva.

VAIOU, D. (1995) Women of the South, after like before Maastricht, in Hadjimichalis, C. and Sadler, D. (eds) *Europe at the Margins: New Mosaics of Inequality*, Wiley, Chichester, pp. 35–49.

VALENTINE, G. (1989) The geography of women's fear, *Area*, **21**, 385–390.

VALENTINE, G. (1993) (Hetero)sexing space: lesbian perceptions and experiences of everyday spaces, *Environment and Planning D: Society and Space*, **11**, 395–413.

VALENTINE, G. (1996) An equal place to work? Anti-lesbian discrimination and sexual citizenship in the European Union, in García-Ramon, M.D. and Monk, J. (eds) *Women of the European Union: the Politics and Work of Daily Life*, Routledge, London, pp. 111–125.

VALENTINE, G. (1997a) 'My son's a bit dizzy. My wife's a bit soft': gender, children and cultures of parenting, *Gender, Place and Culture*, **4**, 37–62.

VALENTINE, G. (1997b) 'Oh yes I can, oh no you can't': children and parents' understandings of kids' competence to negotiate public space safely, *Antipode*, **29**, 65–89.

VALENTINE, G. (1998) Public/private voices 'sticks and stones may break my bones': a personal geography of harassment, *Antipode*, **30**, 205–332.

VARGAS, V. (1991) The women's movement in Peru: streams, spaces and knots, *European Review of Latin American and Caribbean Studies*, **50**, 7–47.

VARLEY, A. (1996) Woman heading households: some more equal than others?, *World Development*, **24**, 505–520.

VERTOVEC, S. (1993) Local context and the development of Muslim communities in Britain: observations in Keighley, West Yorkshire, unpublished paper available from the author.

WALTER, N. (1998) *The New Feminism*, Little Brown, London.

WARE, V. (1992) *Beyond the Pale: White Women, Racism and History*, Verso, London.

WARNER, M. (1994) *Managing monsters: six myths of our time, The 1994 Reith Lectures*, Vintage, London.

WATSON, J.L. (1977) *Between Two Cultures: Migrants and Minorities in Britain*, Basil Blackwell, Oxford.

WATSON, P. (1997) Civil society and the politics of difference in eastern Europe, in Scott, J.W., Kaplan, C. and Keates, D. (eds) *Transitions, Environments, Translations: Feminisms in International Politics*, Routledge, New York, pp. 21–29.

WATTS, H.D., SMITHSON, P.A. and WHITE, P.E. (1989) *Sheffield Today*, Department of Geography, University of Sheffield, Sheffield.

WEICHERT, B. and HÖPFNER, H. (1990) Frauen in Politik und Gesellschaft, in Winkler, G. (ed.) *Frauenreport '90*, Verlag Die Wirtschaft, Berlin, pp. 199–227.

WEINER, G. (ed.) (1985) *Just a Bunch of Girls*, Open University Press, Milton Keynes.

WEISMANTEL, M.J. (1988) *Food, Gender and Poverty in the Ecuadorian Andes*, University of Pennsylvania Press, Philadelphia, PA.

WERBNER, P. (1996) Public spaces, political voices: gender, feminism and aspects of British Muslim participation in the public sphere, in Shadid, W.A.R. and van Koningsveld, P.S. (eds) *Political Participation and Identities of Muslims in Non-Muslim States*, Kok Pharos, Kampen, pp. 53–70.

WGSG (Women and Geography Study Group of the Institute of British Geographers) (1984) *Geography and Gender*, Hutchinson, London.

WGSG (Women and Geography Study Group of the Royal Geographical Society with the Institute of British Geographers) (1997) *Feminist Geographies: Explorations in Diversity and Difference*, Longman, London.

WIEST, K. (1997) *Die Neubewertung Leipziger Altbauquartiere und Veränderungen des Wohnmilieus*, Beiträge zur regionalen Geographie 43, Institut für Länderkunde Leipzig, Leipzig.

WILKINSON, H. and SIANN, G. (1995) *Gender, Feminism and the Future*, Demos, London.

WILSON, A. (1978) *Finding a Voice*, Virago, London.

YOUNG, I.M. (1990a) *Justice and the Politics of Differences*, Princeton University Press, Princeton, NJ.

YOUNG, I.M. (1990b) The ideal of community and the politics of difference, in Nicholson, L.J. (ed.) *Feminism/Postmodernism*, Routledge, London, pp. 300–323.

YUVAL DAVIES, N. (1997) *Gender and Nation*, Sage, London.

Index

Activism, 54, 110, 113–134, 160, 165, 171, 177, 184, 191, 194
Andahuaylas, 48, 62, 67, 74, 79, 82, 83, 85, 87, 88, 160, 172, 177
Asian, 19, 24, 46, 47, 49, 50, 135–151, 180, 183, 184, 190, 194

Binary/ies, 31, 53, 67, 75, 77, 78, 145, 149, 151, 180, 181
Body, 3, 4, 7, 16, 27, 31
Breadwinner, 4, 8, 156, 162
Britain, 1, 7, 8, 13, 14, 15, 23, 28, 32, 35, 44, 91–112, 154, 155, 156, 158, 163, 166, 180, 190

Capitalism, 18, 21, 26, 30, 174, 176, 182, 187
Chatterbox, 105, 110, 171
Child/children, 1, 8, 9, 15, 36, 43, 44, 55, 60, 71, 73, 74–88, 91–112, 114–118, 124, 125, 127–134, 139, 150, 160–169, 177, 179, 190, 193–195
Childcare, 8, 37, 43, 44, 48, 55, 56, 60, 65, 80, 93–112, 114, 133, 161, 164, 165–168, 170, 171, 179
  childminder/s, 93, 101, 102, 110, 111, 168
  family care, 101
  nanny/ies, 93, 111
  local childcare cultures, 55–56, 60, 96, 108–109, 112
  private nursery/ies (commercial nurseries), 93, 101–103, 107
Citizenship, 15, 24, 25, 37, 172, 191, 192

Civil war, Peru, 58, 71, 157, 160
Class, 4–6, 10, 19, 23–33, 36, 39, 40, 49, 52, 53, 56, 57, 61, 82, 92–94, 106, 111, 123, 135, 137, 141, 143, 148–151, 161–167, 170, 171, 183–185, 188, 193–195
clothes (dress), 79, 140, 141, 144, 145, 147–149, 165–168
Community/communities, 10, 13, 19, 33, 36, 37, 44, 58, 64, 67, 68, 72, 75, 81, 82, 89, 90, 110, 111, 118, 129, 136, 141, 151, 154, 160, 170–172, 180, 181, 184
Crime, 179
Cultural politics of difference, 26, 27
Culture (cultural), 1, 3, 6, 8, 10, 12, 14, 16–40, 44, 49, 53, 54, 56, 62, 69, 70, 76, 96, 97, 99, 104, 106, 109, 110, 113, 125, 128, 136–152, 153–184, 188–197

Debt repayment, 68, 80, 155
Diaspora, 22, 23, 35, 179, 180
Difference, 3, 11–13, 21–33, 40, 48–53, 57–58, 101, 103, 140, 152–154, 160–162, 173–185, 187, 191, 194, 197
Discourse, 10, 21, 30, 46, 140, 141, 142, 147, 174
Domestic/domesticity, 6, 7, 10, 12, 14, 19, 25, 33, 35, 39, 43, 62, 70, 71, 72, 74, 78, 83, 85, 86, 87, 111, 133, 139, 140, 157, 162, 163, 164, 165, 168, 182

Domestic violence, 39, 72, 168
Double burden, 9, 37, 115

Economic crisis, 68, 69, 70, 71, 79, 80,
    155, 160, 163, 180, 193
Economic restructuring, 12, 32, 37, 67,
    94, 108, 131, 156, 177, 191, 192,
    194
EEFSU (Eastern Europe and Former
    Soviet Union), 19, 20, 21, 36, 37,
    113–134, 178
Emergency work, 43, 48, 67–90, 155
Epistemology, 41, 42, 196
essentialist/essentialism, 26, 27, 55, 64
Ethnicity/ethnicity, 27, 28, 29, 57, 61,
    135, 137, 161, 183, 184, 188
    white, 'white' (whiteness), 6, 10, 11,
        23, 28, 30, 39, 49, 56, 58, 92, 96,
        137, 138, 140, 143, 149, 156, 197
    black, 13, 28, 49, 149, 197

Father/fatherhood/fathering/fathers, 77,
    91, 93, 96, 99, 105, 143, 166
Femininity
    'appropriate' femininities, 9, 10, 105,
        110, 137, 139, 140, 141, 142, 143,
        144, 147, 150, 151, 165
    'hegemonic' femininities, 4, 28, 188,
        189, 192
Feminisation of poverty, 188
Feminism, 1, 8, 38, 39, 43, 54, 58, 182,
    195
    'popular feminism', 39, 43, 182
Feminist geography, 3, 11, 26, 30, 31,
    41–42, 48, 54
Feminist research, 31, 42, 48–49,
    53–58, 65
'field', 14, 21, 41, 47–48, 52, 58, 120
'fieldwork', 41–42, 45–48, 51–52, 76,
    94, 105, 112, 160
Food, 23, 71, 83–88
Fractured identities, 153, 161, 183

Gender
    gender and development (GAD), 34
    gender division/s, 3, 12
    gender identity/ies, 1–15, 16, 21,
        31–40, 53–54, 62, 65, 74–75, 90,
        108, 111–112, 154–168, 174–186,
        188–196

gender ideologies, 36, 157–158
gender role/s, 11, 37, 39, 58, 62
Gender regime, 5, 36–37, 156–158, 180
Gender relations, 4, 11–12, 30, 34–35,
    39, 43, 58, 59, 62, 137, 153, 156,
    174–177, 181, 185, 189, 194–196
Germany
    Reunification, 9, 15, 37, 43, 45, 52,
        54, 64, 113–114, 116–117, 121,
        124–125, 127, 130, 134, 138, 157,
        164, 180, 189
    East Germany, 157, 188
    West Germany, 7, 64, 155
Global/local, 13, 153, 173
Globalisation, 3, 16–26, 32–40, 67,
    113, 153, 156, 174–185, 189,
    194–196

Hallam, 94–112, 191
headscarf, 58, 143–149
Health(care), 19, 33–34, 69, 71,
    103–106, 108, 110
Heteropatriarchy, 13
Heterosexual, 9, 30, 35, 76, 92, 156
Heterosexuality, 1, 4, 28, 30
hijab (see headscarf)
Home, 5–8, 12–15, 30, 37, 49, 52–55,
    70–72, 83, 86–88, 92, 97–105,
    108–112, 114–117, 127–134,
    136–151, 154, 161–173, 180–185,
    191–195
Homosexuality, 27, 76
Household
    female heads of household, 70, 73,
        75, 179
    household income, 9, 93, 94, 151,
        163
Human rights, 71

Identity/ies, 3–7, 11, 14, 23–40, 43–50,
    53–57, 60, 65, 97, 127, 132–134,
    135–139, 141, 146–151, 160, 162,
    164, 166, 171–186, 190–197
    hybrid identity/ies, 136, 184
Income (Money), 9, 23, 37, 43, 67–76,
    79, 80, 83, 85, 88–89, 93–94,
    97–99, 106–108, 115, 150,
    163–167, 177, 191–193
Interview/s, 44–47, 50–51, 60, 63, 76,
    97, 118, 138, 143

Isolation, 14, 16, 25, 97, 106, 127, 168, 183, 195

Justice, 44, 55, 60

Labour
  flexible labour, 70
  green labour, 12, 74
  labour force, 7, 12, 18, 32–34, 70, 74, 108, 115, 132, 164, 187
Labour market, 18, 21–22, 33, 94, 97, 99–100, 112, 115–116, 130, 132, 161, 165, 193
lesbian/s, 1, 13, 27–28, 76, 181
Liberation, 22, 181
Lima, 43, 48, 51, 62, 67–89, 160, 171, 177
local (locality), 12–18, 21–26, 32–37, 45–46, 51–52, 55–56, 60, 64–65, 70–72, 81–82, 90, 93–97, 104–112, 113–114, 117–134, 141, 153–154, 158, 161, 167, 170–179, 184, 189, 191, 194, 196
London, 13, 15, 18, 33, 36, 48, 135, 138, 156

Marianismo, 10, 70, 157
Marriage, 8, 93, 142, 150, 167
Masculinity/ies, 3–5, 11, 14, 27–28, 31, 133, 158, 162–163, 166, 169, 181, 187
Maternal responsibility, 98–99, 177, 191
Methodology, 14, 41–66, 94, 197
Methods, 41–45, 73
Migration, 22, 34, 153, 156
Moral geographies, 13, 56, 109–110, 112, 171, 192, 196
Moral geographies of mothering, 109–112, 171
Mother, 4, 8, 10, 56, 75–80, 88, 97–102, 104–111, 132, 134, 139, 142–143, 148, 157, 166, 169, 171, 184, 199
Motherhood, 7, 13–15, 36–39, 56, 67, 70, 78, 91–92, 96, 100, 108–112, 115, 143, 150, 162, 166
Mothering, 5, 15, 36, 44, 67, 80, 83, 91–99, 108–112, 129, 132, 166, 169, 171–172, 179, 191–193

Muslim, 5, 15, 36, 43, 46–47, 49–50, 56–58, 63, 135–152, 156, 158, 166–170, 172, 180, 183–184, 190, 194

Nation
  nation state, 23, 37, 180
  nationhood, 5, 23
  nationalism, 10, 21, 23, 39
Negotiation, 10, 13, 15, 32, 35, 47, 56, 67, 75, 83, 86, 90, 126, 138, 146, 149–150, 168, 185, 188–189, 194, 196
Neighbourhood, 13, 15, 37, 39, 48, 64, 76, 110–112, 113–114, 117–129, 132, 134, 138, 141, 161–162, 164–172, 181–185, 191–196
Neo–liberal/ism, 3–5, 8, 10, 16–18, 19–21, 25–26, 34, 36–37, 113, 153, 155, 174, 175, 177–180, 185, 189, 195
network/s, 12–13, 18, 20, 23–26, 29, 33, 35, 37, 51, 81, 96–97, 101, 104–106, 110–112, 154, 157, 161–162, 169–173, 182–185, 193, 196
New International Division of Labour (NIDL), 33
New Social Movements (NSMs), 25–26, 28–29, 38–39, 43, 71, 177
New World Order, 3, 24
NGOs, 19, 157, 182

PAIT, 68–89, 155, 160, 164–165, 168, 171–172, 182, 184, 194
Pakistan, 138, 143, 150, 180
patriarchy (patriarchal), 7, 30–37, 58, 115–116, 137, 142, 152, 157, 164, 180, 182, 194
Performance, 4, 28, 31, 53, 76, 115, 149, 181, 187, 193
Peru, 5, 10, 14, 23, 34, 36, 43, 48, 51, 58, 67–71, 91, 154–158, 160, 163–165, 169, 171–172, 177, 180, 182–183, 189, 193
Place, 12–15, 18, 21, 27, 33, 40, 44, 47–48, 59–65, 72–82, 86–87, 101–102, 106, 109–112, 120, 123, 128, 135, 138, 148, 154, 156, 158, 160–162, 165–166, 169, 172–176, 181–185, 189–195, 197

Politics
  gender politics, 3, 89
Popular feminism/s, 39, 43, 182
Positionality, 14, 42, 47–53, 59, 65, 196
Poverty, 1, 68, 71, 100, 188
Power relations, 22, 25, 41–42, 48–50, 52, 54, 58, 111, 113, 187
Pre–school education
  Playgroups, 93, 103, 105, 110, 171, 177, 181
  LEA nursery/ies, 103, 106–107

Qualitative methods, 41, 45
  Observation, 45–46, 86, 138
  archives, 44
Quantitative methods, 42, 45
  questionnaire survey, 44–45, 60, 94, 96, 121, 128

Race, 4, 9–10, 19, 29, 40, 47, 49–50, 135, 162
  Racialised, 4, 10, 16, 27, 29, 31, 35, 53, 137, 139, 142–143, 154, 183
Racism, 13, 31, 47, 50, 86, 137, 183, 194
Reflexivity, 52, 65
Religion, 10, 47, 57, 137, 139–140, 144, 146, 183
Representation/s, 10, 23, 33, 42, 49–50, 59, 65, 80, 90, 136, 158, 183, 192, 194, 195, 197
Research subjects, 53, 56–57
Resistance, 24–25, 54–55, 58, 147, 162, 195

School, 1, 15, 34, 43–44, 47–48, 51, 55, 57, 63, 69, 72–73, 82, 92–112, 114, 118, 128, 133, 135–151, 161, 165, 167–169, 172, 184, 190, 192, 194, 196
sexuality/sexualities, 3–4, 9, 19, 27–28, 39–40, 55–56, 137, 140–142, 144, 147–149, 161–162, 178, 183–184, 188
Sheffield, 44, 48, 94, 100, 106, 112, 156, 158, 161, 163, 165, 170–171, 183–184, 191–193
Situated knowledges, 50

Social movements, 25–26, 28, 39, 43, 67, 71, 192
Social networks (see network)
Social security, 68, 69
Southey Green, 94, 96–112, 177, 191
Space, 6, 12–15, 18, 21–22, 24–25, 32–33, 35, 39–40, 47–48, 51–52, 54–55, 59–60, 63–67, 69, 71–76, 80–83, 87–89, 91, 105, 110–111, 113, 117–118, 121–122, 128–136, 141, 145, 148–154, 162, 166–186, 188–190, 193–197
  everyday spaces, 59, 63, 127, 154, 161–162, 173, 176, 185, 195–196
  imagined space, 136, 149, 184
  local space, 80–82, 113, 117, 169, 171, 176
  private space/sphere, 39, 62, 87
  public space/sphere, 14, 64, 69, 71, 80–83, 87, 118, 148, 151, 154, 167, 160, 161
  relational space, 59, 71
  'third space', 136
Spice Girls, 1, 135, 187
Structural adjustment, 155, 178
Survival strategies, 43, 79, 157, 177, 193–194
Syncretism, 180

Tank Girl, 2
Third World, 19, 25, 31–34, 51, 176–179, 182
Toddler groups, 104–105, 110
Transgression, 24, 182, 187, 195
Transition, 5, 15, 20–21, 37, 45, 54–55, 63–64, 113, 116, 118, 123, 132, 134, 150, 153, 164, 167, 174–175, 181, 185, 189, 195
Transnational, 19, 22–24, 26, 157–158

Underemployment, 69–70
Unemployment, 20, 37, 53–69, 94, 99, 108, 116, 130–131, 150, 155–156, 158, 162, 164, 167, 179

veil (see headscraf)

Welfare, 3, 25, 37, 67–68, 80, 90, 128, 130, 155–157, 160, 178–179

Western, 8–9, 19–21, 30, 37, 49, 51, 54–55, 58, 63–64, 94, 106, 114, 116, 118, 120–123, 126, 132–134, 137, 144–145, 148–151, 155, 157, 164, 166–167, 169, 171, 175–179, 188–191
Wives, 8, 37, 75–79, 139, 142, 165–166
  Housewives, 8, 75, 82, 156, 167, 189, 191
Women's movement, 26

Work
  paid work, 15, 18, 43, 51, 62, 73–74, 78–79, 83, 93, 97–98, 107, 110–111, 115, 127, 133, 154, 161–165, 168–169, 181–185, 190–195
Workfare, 69–90, 116, 165, 172, 177, 190
Workplace/workspace, 7, 13, 78, 83, 86, 88, 97, 115, 127, 131, 163, 196